T0320413

REVERSIBLE WORLD OF CELLULAR AUTOMATA

Fantastic Phenomena and Computing in Artificial Reversible Universe

WSPC Book Series in Unconventional Computing

Print ISSN: 2737-5218
Online ISSN: 2737-520X

Published

Vol. 6　*Advances in Quantum Computer Music*
　　　　edited by Eduardo Reck Miranda

Vol. 5　*Post-Apocalyptic Computing*
　　　　edited by Andrew Adamatzky

Vol. 4　*Reversible World of Cellular Automata: Fantastic Phenomena and*
　　　　Computing in Artificial Reversible Universe
　　　　by Kenichi Morita

Vol. 3　*Actin Computation: Unlocking the Potential of Actin Filaments for*
　　　　Revolutionary Computing Systems
　　　　edited by Andrew Adamatzky

Vol. 2　*Unconventional Computing, Arts, Philosophy*
　　　　edited by Andrew Adamatzky

Vol. 1　*Handbook of Unconventional Computing*
　　　　(In 2 Volumes)
　　　　Volume 1: Theory
　　　　Volume 2: Implementation
　　　　edited by Andrew Adamatzky

WSPC BOOK SERIES IN UNCONVENTIONAL COMPUTING   *Volume 4*

REVERSIBLE WORLD OF CELLULAR AUTOMATA

Fantastic Phenomena and Computing in Artificial Reversible Universe

Kenichi Morita

Hiroshima University, Japan

World Scientific

NEW JERSEY · LONDON · SINGAPORE · BEIJING · SHANGHAI · HONG KONG · TAIPEI · CHENNAI · TOKYO

Published by

World Scientific Publishing Co. Pte. Ltd.

5 Toh Tuck Link, Singapore 596224

USA office: 27 Warren Street, Suite 401-402, Hackensack, NJ 07601

UK office: 57 Shelton Street, Covent Garden, London WC2H 9HE

Library of Congress Cataloging-in-Publication Data
Names: Morita, Ken'ichi (Researcher in natural computation), author.
Title: Reversible world of cellular automata : fantastic phenomena and computing in
    artificial reversible universe / Kenichi Morita, Hiroshima University, Japan.
Description: New Jersey : World Scientific, [2025] |
    Series: WSPC book series in unconventional computing, 2737-5218 ; volume 4 |
    Includes bibliographical references and index.
Identifiers: LCCN 2024033289 | ISBN 9789811280320 (hardcover) |
    ISBN 9789811280337 (ebook for institutions) | ISBN 9789811280344 (ebook for individuals)
Subjects: LCSH: Cellular automata. | Reversible computing.
Classification: LCC QA267.5.C45 M67 2025 | DDC 006.3/822--dc23/eng/20240904
LC record available at https://lccn.loc.gov/2024033289

British Library Cataloguing-in-Publication Data
A catalogue record for this book is available from the British Library.

For any available supplementary material, please visit
https://www.worldscientific.com/worldscibooks/10.1142/13516#t=suppl

Desk Editors: Soundararajan Raghuraman/Steven Patt

Typeset by Stallion Press
Email: enquiries@stallionpress.com

Printed in Singapore

To my parents, my wife, my children, and my grandchildren

Preface

This is a puzzle book. It proposes various mathematical puzzles on how we can find fascinating phenomena in a space that obeys a very simple reversible law, and how we can create objects that perform amazing tasks in it. To our scientific knowledge, our physical world is governed by a reversible law in the microscopic level, as it is seen in quantum mechanics. Therefore, solving the above kind of puzzles will help us to understand and utilize microscopic properties of our real world. In particular, when we try to construct future computing devices that directly use microscopic physical phenomena, it is necessary to consider the reversibility, one of the fundamental laws of nature.

A cellular automaton (CA) is an abstract spatiotemporal model of dynamical systems. In this book, we use a 2-dimensional reversible cellular automaton (RCA) as a model of a reversible world. Although it is an artificial universe, it is useful for studying which kinds of phenomena emerge from a simple reversible law. Here, we use the framework of partitioned cellular automata (PCA), in which each cell is divided into several parts and changes its state depending on the neighboring cells' parts. By this, we can design reversible CAs very easily. Subclasses of PCAs called elementary square PCAs (ESPCAs) and elementary triangular PCAs (ETPCAs), which have very simple local transition rules, play crucial roles in this book. Despite the extreme simplicity of the reversible elementary PCAs, they show quite interesting behavior. Thus, there arise various puzzle problems. The most important ones are the following: Which reversible elementary PCA is computationally universal, and how can we compose universal computers in its cellular space?

In this book, attractive puzzle problems like the above are posed and solved. Though solutions for many of these puzzles are given here, readers are recommended to consider and solve them by themselves, and further propose new interesting puzzle problems. Most puzzles are solved by giving configurations of reversible PCAs, which are the states of the whole cellular space. However, in some cases, configurations given as solutions are very complex, and need millions of cells. Such a case occurs, *e.g.*, when composing configurations that simulate universal computers. To verify the correctness of these solutions, a good simulator for CAs is absolutely necessary besides a theoretical method. Here, we use *Golly* (`https://golly.sourceforge.io/`), a general purpose CA simulator, which can simulate evolving processes of very large configurations for a huge number (say, millions or even billions) of steps. Solutions for many puzzle problems are given as rule files and pattern files that are executable on Golly. They are available at `https://ir.lib.hiroshima-u.ac.jp/ 00055227`.

There is no prerequisite knowledge to read this book, because the framework of the CAs is very simple. Since there are many (*i.e.*, about 370) figures, it may be good to look at only them at first. By this, readers can feel the atmosphere of the whole volume. Though it is better to have some basic knowledge on automata theory to understand several particular sections precisely, it is not required to do so to grasp the main objective of the book. Instead, I strongly recommend the readers to use the simulator Golly, and see fantastic evolving processes of various reversible CAs. By this, the readers can see the visualized forms of the solutions for the puzzles. I hope the book and the files for Golly will open a new vista for a reversible world of cellular automata.

Kenichi Morita
May 2024

About the Author

 **Kenichi Morita** is a professor emeritus of Hiroshima University. He received his B.Eng., M.Eng., and Ph.D. degrees from Osaka University in 1971, 1973, and 1978, respectively. From 1974 to 1987, he was a research associate at the Faculty of Engineering Science, Osaka University. From 1987 to 1990, he was an associate professor, and from 1990 to 1993, a full professor at the Faculty of Engineering, Yamagata University. From 1993 to 2013, he was a full professor at the Graduate School of Engineering, Hiroshima University. His research interests include automata theory, reversible computing, and formal language theory. In these research subjects, he has engaged in the investigation of theory of reversible computing and reversible cellular automata (CAs) for more than 35 years and published books and many papers on them. In particular, a systematic and comprehensive study on reversible computing is given in the book *Theory of Reversible Computing* (Springer, 2017). Some of the important results shown in these books and papers are proposal of the useful framework of partitioned CAs (PCAs) for studying reversible CAs, proving computational universality of one-dimensional reversible CAs and reversible two-counter machines, proposal of simple reversible logic elements with one-bit memory (RLEMs) and clarifying their universality, showing that even extremely simple two-dimensional reversible CAs can simulate reversible Turing machines, and so on.

Acknowledgments

First, I express my thanks to the developing team of Golly. It is definitely an excellent CA simulator. Without Golly, the work given in this book has not been accomplished. I started the studies on reversible computing and reversible CAs around the middle of 1980s in Osaka University. I continued the studies in Yamagata University, and then in Hiroshima University. There, the former colleagues and students greatly helped me to advance the studies. It was indeed my pleasure to work with them. I express my special gratitude to all of them. I also express my great thanks to many researchers I met in many places in the world for their helpful discussions and encouragement.

Contents

Fantastic Phenomena and Computing in Reversible Cellular Automata

145

Chapter 1

Cellular Automaton as an Artificial Digital World

Cellular automata (CAs) are discrete systems consisting of a large number of simple elements called cells, each of which changes its state by interacting with its neighbor cells. They can describe various spatiotemporal processes. It is really inspiring to use them for playing, designing, and creating. They often show fantastic phenomena even from very simple interaction rules. It is a fun to play with them and to watch their evolution processes. Once you have played with them, you will have a question what will happen if they start from a different spatial configuration, or if they have different interaction rules. From this point, the designing phase of CAs starts. Designing initial configurations and interaction rules for CAs is just designing useful algorithms or programs. In fact, if we do so suitably, we can create marvelous phenomena, useful functional objects, and even universal computing systems in them. In this way, CAs can be used as a world for creation.

Reversible cellular automata (RCAs) are systems that reflect physical reversibility. Since reversibility is one of the fundamental microscopic laws of nature, it is important to study how interesting phenomena emerge from simple reversible laws. RCAs are abstract models of reversible worlds for studying such problems. As we shall see in the following chapters, RCAs show a high capability of information processing even if they have quite simple reversible local transition rules.

In this chapter, first, CAs and RCAs are briefly introduced. Second, objectives of studying RCAs in this book are given. Third, a general purpose CA simulator *Golly* [1] is explained, since it is used for designing RCAs and for creating interesting objects given in this book. Actually, it is a very useful tool for readers to understand evolution processes of RCAs. Fourth, how to read this book is described. Finally, mathematical terminology and notations are summarized.

1.1 What Are Cellular Automata?

A cellular automaton was first proposed by John von Neumann in 1950's [2] based on the suggestion of Stanislaw Ulam [3] to design and create a *self-reproducing automaton*. In his 2-dimensional CA, each cell has 29 kinds of states. Using an ingenious design method, he created a pattern composed of the 29 states that has a function of a Turing machine and reproduces itself in the cellular space just like living things.

After the pioneering work of von Neumann, various researches on CAs appeared. In them, the *Game of Life* (GoL) proposed by John Horton Conway is the most famous CA. It was first introduced by Martin Gardner in his column of *Scientific American* [4,5]. It attracted not only researchers of CAs but also many amateur scientists, and thus made a great influence on the study of CAs. The GoL's wonder-world can be seen in [1, 6–8]. Actually, the study presented in this book inherits the spirit of GoL.

1.1.1 *Definition and examples of cellular automata*

We first give a definition of a CA. However, readers can understand the example CAs given in this section by referring only their figures.

Definition 1.1. A *k-dimensional cellular automaton* (CA) is defined by

$$A = (\mathbb{Z}^k, Q, (\mathbf{n}_1, \ldots, \mathbf{n}_m), f, \#)$$

Here, $\mathbb{Z}^k$ is the k-dimensional *cellular space*. It is the set of all integer coordinates where *cells* are placed. The set Q is a non-empty finite set of cell's states. The m-tuple $(\mathbf{n}_1, \ldots, \mathbf{n}_m) \in (\mathbb{Z}^k)^m$ is a *neighborhood* ($m \in \{1, 2, \ldots\}$), and thus A is called an *m-neighbor CA*. Each $\mathbf{n}_i$ ($i \in \{1, \ldots, m\}$) is a relative coordinates of a neighbor cell from the cell under consideration. The item $f : Q^m \to Q$ is a *local function*, which determines the next state of each cell from the present states of m neighboring cells. The state $\# \in Q$ is a *quiescent state* (*i.e.*, a blank state) that satisfies $f(\#, \ldots, \#) = \#$.

A *configuration* of the CA A is a mapping $\alpha : \mathbb{Z}^k \to Q$, which gives a whole state of the cellular space $\mathbb{Z}^k$. The set of all configurations of A is denoted by $\mathrm{Conf}(A)$. Applying the local function f to all the cells simultaneously, we obtain a *global function* $F : \mathrm{Conf}(A) \to \mathrm{Conf}(A)$. Applying the global function F repeatedly, configurations of A evolve. Here, we do not give the precise definition of F (see, *e.g.*, [9] for its definition). Note that for partitioned cellular automata (PCAs) introduced in Chap. 2, global functions are defined formally (Definitions 2.5 and 2.21).

In the case of 2-dimensional CA, there are two kinds of commonly used neighborhoods. The first one is:

$$N_N = ((0,0),(0,1),(1,0),(0,-1),(-1,0))$$

It is called the *von Neumann neighborhood*, since it was used in [2]. Each element of N_N gives the relative coordinates from the *center cell* having the relative coordinates $(0,0)$. Thus, the von Neumann neighborhood consists of the following five cells: the center cell and the four adjacent cells having a common edge with the center cell. It is shown in Fig. 1.1(a).

The second one is:

$$N_M = ((0,0),(0,1),(1,1),(1,0),(1,-1),(0,-1),(-1,-1),(-1,0),(-1,1))$$

It is called the *Moore neighborhood*, since it was used in [10]. The Moore neighborhood consists of nine cells, which are the center cell and the eight adjacent cells having either a common edge or a common corner with the center cell as shown in Fig. 1.1(b).

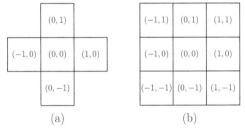

Fig. 1.1   (a) The von Neumann neighborhood, and (b) the Moore neighborhood. Each coordinates shows the relative coordinates from the center cell.

Consider the case where a CA A has the von Neumann neighborhood. For each $q_0, q_1, q_2, q_3, q_4, q \in Q$, if $f(q_0, q_1, q_2, q_3, q_4) = q$ holds, this relation is called a *local transition rule* of A. It can be depicted as shown in Fig. 1.2. The local function f is thus defined as a set of local transition rules. The case of the Moore neighborhood is similar.

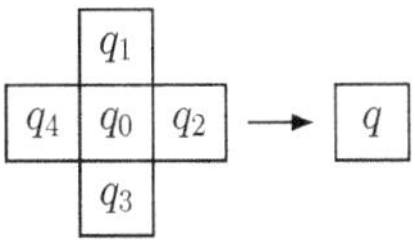

Fig. 1.2   Pictorial representation of the local transition rule $f(q_0, q_1, q_2, q_3, q_4) = q$ of a 2-dimensional CA with the von Neumann neighborhood.

We now give two examples of 2-dimensional CAs. They are the Fredkin's self-replicating CA, and the Game of Life.

Example 1.1 (Fredkin's self-replicating CA). It is a 2-state CA with the von Neumann neighborhood defined below, which was introduced in [5].

$$A_F = (\mathbb{Z}^2, \{0, 1\}, N_N, f_F, 0)$$

Here, the local function f_F is defined by the 16 local transition rules shown in Fig. 1.3. Note that, in this CA, the next state does not depend on the current state of the center cell. Thus, it is actually a 4-neighbor CA.

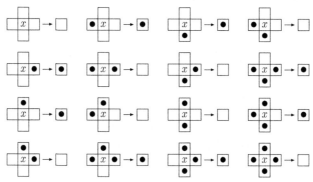

Fig. 1.3   Local transition rules that define the local function f_F of the Fredkin's CA. The states 0 and 1 are represented by a blank and a particle. Here, $x \in \{0, 1\}$.

From Fig. 1.3, we can see that the next state of the center cell becomes 1, if and only if the total number of 1's contained in the current states of the four adjacent cells is odd. Therefore, the local function f_F is also expressed by the following formula, where $\oplus$ denotes the mod 2 addition. Thus, it is a linear CA discussed in Sec. 5.6.1.

$$\forall q_0, q_1, q_2, q_3, q_4 \in \{0, 1\} : f_F(q_0, q_1, q_2, q_3, q_4) = q_1 \oplus q_2 \oplus q_3 \oplus q_4$$

Figures 1.4 and 1.5 show evolution processes of configurations of A_F. We can see that *any* initial pattern given at $t = 0$ replicates. This fact can be proved using the property of the mod 2 addition (Exercise 1.2).

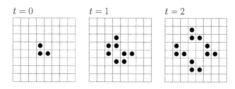

Fig. 1.4   Evolution process in the Fredkin's self-replicating CA [5].

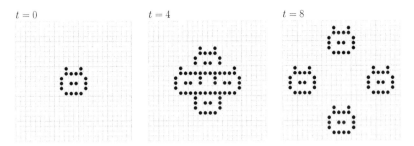

Fig. 1.5   Another evolution process in the Fredkin's self-replicating CA.

Note that the self-replication in the Fredkin's CA looks like a crystal growth rather than the self-reproduction in living things. In fact, such phenomena are caused by the local function of the CA, not by the property of the pattern. On the other hand, in the 29-state CA of von Neumann [2], reproduction of a machine is carried out by manipulating the "gene" of its body by a program stored in the machine. Hence, it is very similar to the biological self-reproduction that uses DNA. See the discussion by Langton [11] on the criterion for the true self-reproduction of machines.

Example 1.2 (Game of Life (GoL)). It is a 2-state CA having the Moore neighborhood [4–6]. It is given as follows.

$$A_{\text{GoL}} = (\mathbb{Z}^2, \{0,1\}, N_{\text{M}}, f_{\text{GoL}}, 0)$$

The states 0 and 1 stand for the dead and live states, and are indicated by a blank and a particle, respectively. Since A_{GoL} has the Moore neighborhood, we need quite many local transition rules to describe f_{GoL}, even if we use the facts that f_{GoL} is rotation-symmetric (*i.e.*, for any local transition rule, there is one obtained by rotating the left-hand side by 90°), and reflection-symmetric (*i.e.*, for any local transition rule, there is one obtained by taking a mirror image of the left-hand side). Fortunately, it is expressed very shortly as follows [6].

"Just 3 for BIRTH, 2 or 3 for SURVIVAL"

It says that a dead cell will become live at the next time step if and only if the number of live cells among the eight neighboring cells is just 3, and a live cell will keep live if and only if the number of live cells among the eight neighboring cells is 2 or 3. Thus, f_{GoL} is expressed by the following formula.

$$\forall q_0, q_1, \ldots, q_8 \in \{0, 1\}:$$
$$f_{\text{GoL}}(q_0, q_1, \ldots, q_8) = \begin{cases} 1 & \text{if } q_0 = 0 \ \land \ (\sum_{i=1}^{8} q_i) \in \{3\} \\ 1 & \text{if } q_0 = 1 \ \land \ (\sum_{i=1}^{8} q_i) \in \{2, 3\} \\ 0 & \text{elsewhere} \end{cases}$$

Figure 1.6 shows an example of an evolution process of configurations in GoL. Each of the patterns at $t = 0, 1$ and 2 consisting of three live cells is called a *blinker* [4]. At $t = 0$ the three live cells are aligned vertically. At $t = 1$ they are aligned horizontally, and at $t = 2$ the pattern becomes the same as the one at $t = 0$. Hence, it is a periodic pattern.

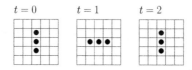

Fig. 1.6   Evolution process of a *blinker* in GoL [4].

Figure 1.7 shows another example. At $t = 0$, a *Cheshire Cat* pattern [5] is given. Note that Cheshire Cat is a cat that appears and disappears in *Alice's Adventures in Wonderland* [12]. Further note that the introduction and the notes in the book [12] were written by Martin Gardner. In [5] it is written that the Cheshire Cat disappears leaving a paw print at $t = 7$ as shown in Fig. 1.7. Since the pattern at $t = 7$ is a stable one called a *block*, it remains the same for $t \geq 7$.

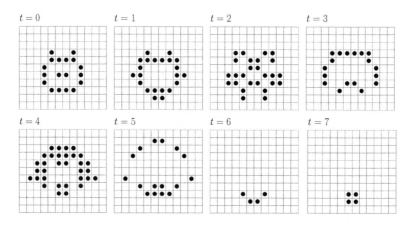

Fig. 1.7   Another evolution process in GoL [5]. It starts from a configuration having the *Cheshire Cat* pattern ($t = 0$). It finally leaves a stable pattern *block* at $t = 7$.

In Sec. 5.1, we consider GoL again. There, similarities and differences of the properties between GoL and reversible CAs are discussed.

1.1.2 *Reversible cellular automata*

We define a reversible cellular automaton, and explain how to design it.

1.1.2.1 *Defining reversible cellular automata*

We first consider an injective cellular automaton and a surjective one.

Definition 1.2. Let A be a CA whose global function is F. If F is injective, A is called an *injective cellular automaton*. If F is surjective, A is called a *surjective cellular automaton*.

The history of the study of injective and surjective CAs is rather long. Moore [10] and Myhill [13] showed the so-called Garden-of-Eden Theorem in early 1960's. This theorem gives a relation between the injectivity and the surjectivity of a global function. Note that a *Garden-of-Eden configuration* is one such that there is no predecessor configuration of it, and thus it can exists only at time $t = 0$. Therefore, a global function of a CA is surjective, if and only if the CA has no Garden-of-Eden configuration.

A configuration α of a CA $A = (\mathbb{Z}^k, Q, N, f, \#)$ is called *finite* if the set $\{\mathbf{x} \in \mathbb{Z}^k \mid \alpha(\mathbf{x}) \neq \#\}$ is finite (*i.e.*, the number of non-blank cells is finite). Let F be the global function of A, and let F_{fin} denote the restriction of F to the set of all finite configurations. The following theorem is obtained by combining the results in [10] and [13].

Theorem 1.1 (Garden-of Eden theorem [10, 13]). *Let A be a CA and F be its global function. Then, F is surjective if and only if F_{fin} is injective.*

We have the following corollary, since it is clear that if F is injective, then F_{fin} is injective.

Corollary 1.1. *Let A be a CA, and F be its global function. If F is injective, then it is surjective.*

Toffoli [14] studied injective and surjective CAs as models of physically reversible systems, and defined a *reversible CA* as the one having a bijective global function. However, in this book, we use the terms "reversible CA" and "injective CA" synonymously, since Corollary 1.1 holds.

Definition 1.3. If A is an injective CA, it is called a *reversible cellular automaton* (RCA).

There is also a related notion called *invertibility* of a CA.

Definition 1.4. A CA $A = (\mathbb{Z}^k, Q, N, f, \#)$ with the global function F is called an *invertible cellular automaton*, if there exists a CA $A' = (\mathbb{Z}^k, Q, N', f', \#)$ with the global function F' that satisfies the following.

$$\forall \alpha, \beta \in \mathrm{Conf}(A) : \ F(\alpha) = \beta \text{ if and only if } F'(\beta) = \alpha$$

Namely, the CA A' undoes evolution processes of A.

From the above definition, it is clear that if a CA A is invertible, then it is injective. Its converse is derived from the results independently proved by Hedlund [15] and Richardson [16]. Thus, we have the following.

Theorem 1.2 ([15, 16]). *A CA A is injective if and only if it is invertible.*

By above, the notions of reversibility, injectivity, and invertibility are all equivalent. Furthermore, if a CA is reversible, it is surjective.

1.1.2.2 *Designing reversible cellular automata*

Designing a 2-dimensional (or higher-dimensional) reversible CA is difficult, if we use a standard framework of a CA given in Definition 1.1. This is because there is no decision algorithm for injectivity of a 2-dimensional CA. It was proved by Kari [17].

Theorem 1.3 ([17]). *The problem whether the global function of a given 2-dimensional CA is injective or not is undecidable.*

On the other hand, for the case of 1-dimensional CAs, the following result is shown by Amoroso and Patt [18].

Theorem 1.4 ([18]). *There is an algorithm to determine whether the global function of a 1-dimensional CA is injective or not.*

Later, a quadratic time algorithm for 1-dimensional CAs was given by Sutner [19]. However, when we try to design a 1-dimensional reversible CA, using these algorithms is not practical. It is desirable that a much easier method for designing reversible CAs is provided.

Since a local function of a CA $A = (\mathbb{Z}^k, Q, N, f, \#)$ is $f : Q^k \to Q$, f can be injective only if $|Q| = 1$ or $m = 1$. We can see that, if f is injective, then the global function F is also injective. First, in the case of $|Q| = 1$, F is trivially injective.

Second, in the case of $m = 1$, f is injective, if and only if there is a bijection (*i.e.*, permutation) $\pi : Q \to Q$ such that $f(q) = \pi(q)$. If $N = (\mathbf{n}_1)$ ($\mathbf{n}_1 \in \mathbb{Z}^k$), then the global function F satisfies $F(\alpha)(\mathbf{x}) = \pi(\alpha(\mathbf{x} + \mathbf{n}_1))$ for every $\mathbf{x} \in \mathbb{Z}^k$ and for every configuration α of A. Therefore, F is injective, since $F^{-1}(\alpha)(\mathbf{x}) = \pi^{-1}(\alpha(\mathbf{x} - \mathbf{n}_1))$. However, such a reversible CA is not interesting at all, since a cell-by-cell permuted configuration simply appears at the position shifted by $-\mathbf{n}_1$ by an application of F.

However, as we shall see in the following chapters, there are quite many interesting reversible CAs whose local functions are not injective if we use the traditional framework of CAs given in Definition 1.1. Hence, we need a new framework for CAs in which reversible CAs are easily designed, and also these CAs are a subclass of the traditional CAs.

Partitioned cellular automaton (PCA)

For this purpose, we introduce the framework of a *partitioned cellular automaton* (PCA). In an m-neighbor PCA, a cell is divided into m parts, and each part has its own state set. Thus, the set of states of a cell is the Cartesian product of the sets of states of m parts. Figure 1.8(a) and (b) are examples of cellular spaces of 5-neighbor and 4-neighbor square PCAs, respectively, and Fig. 1.8(c) is that of a 3-neighbor triangular PCA.

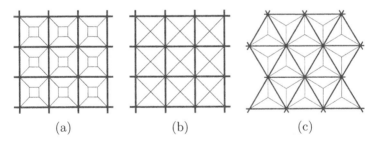

(a)   (b)   (c)

Fig. 1.8   Cellular spaces of (a) 5-neighbor square PCA, (b) 4-neighbor square PCA, and (c) 3-neighbor triangular PCA.

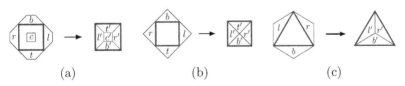

(a)   (b)   (c)

Fig. 1.9   Forms of local transition rules of (a) 5-neighbor square PCA, (b) 4-neighbor square PCA, and (c) 3-neighbor triangular PCA.

Figure 1.9(a) shows the form of a local transition rule for the 5-neighbor square PCA. It describes that the next state (c', t', r', b', l') of the center cell is determined by the present state c of the center part of the center cell, the state t of the top part of the south cell, the state r of the right part of the west cell, the state b of the bottom part of the north cell, and the state l of the left part of the east cell. We can see that it corresponds to the case of the von Neumann neighborhood. However, the next state (c', t', r', b', l') is determined not depending on the whole states of the five neighbor cells.

Figure 1.9(b) is for the 4-neighbor square PCA. It is a special case of the 5-neighbor square PCA where the center part has only one state.

Figure 1.9(c) is for the 3-neighbor triangular PCA. Note that this local transition rule is for the up-triangle cells ($\triangle$). For the down-triangle cells ($\triangledown$), we should use the local transition rule obtained by rotating the both sides of Fig. 1.9(c) by 180°.

An important fact in PCAs is that the domain and the codomain of a local function is always the same. Therefore, it can be injective. As it will be shown in Theorem 2.1, in a 4-neighbor square PCA, its local function is injective if and only if its global function is injective. This property for a 5-neighbor square PCA, and for a 3-neighbor triangular PCA is proved similarly (a proof for a general PCA is found in [9]). By this property, we can give reversible PCAs very easily.

Furthermore, we can see that PCAs are a subclass of traditional CAs. This fact will be shown in Theorem 2.2. Therefore, by designing a reversible PCA that has an injective local function, we obtain a reversible CA. Another advantage of a PCA is that we can define a PCA having an arbitrary neighborhood.

Although there are a few other frameworks for designing reversible CAs, such as block CAs and second-order CAs, we use PCAs in this book. In the following, these CA models are described briefly.

Other models of RCAs

A *block cellular automaton* was proposed by Margolus [20]. In his CA, all the cells are grouped into blocks of size 2×2 as in Fig. 1.10. Its local function maps a block state (*i.e.*, a state of a 2×2 array of cells) into a block state. At time 0 the local function is applied to every solid-line block, then at time 1 to every dashed-line block, and so on, alternately. This kind of neighborhood is called the *Margolus neighborhood*. We can see that if the local function is injective, then its global function is injective. In this way, reversible CAs can be obtained. However, CAs with Margolus neighborhood are not traditional CAs, since each cell must know the relative position in a block and the parity of time besides its own state. A relation between a block CA and a 4-neighbor square PCA will be discussed in Sec. 6.3.

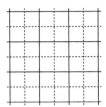

Fig. 1.10   Cellular space of a 2-dimensional CA with the Margolus neighborhood.

A *second-order cellular automaton* is one such that the state of a cell at time $t + 1$ is determined not only by the states of the neighbor cells at time t but also by the state of the center cell at $t - 1$. A *reversible second-order cellular automaton* was proposed by Margolus [20]. He gave a method of using second-order CAs to compose reversible ones, and designed a specific model in which universal reversible logic gate can be implemented.

Let $A = (\mathbb{Z}^k, Q, (\mathbf{n}_1, \dots, \mathbf{n}_m), f)$ be an arbitrary CA. We assume a binary operator $\ominus$ is defined on the state set Q, which satisfies $a \ominus b = c$ if and only if $a \ominus c = b$ for all $a, b, c \in Q$. A typical example is $Q = \{0, 1, \dots, r - 1\}$ and the operation $\ominus$ is the mod r subtraction. Let $\alpha^t \in \mathrm{Conf}(A)$ denote a configuration at time t. We define a kind of global function $\tilde{F} : (\mathrm{Conf}(A))^2 \to \mathrm{Conf}(A)$, which determines the configuration α^{t+1} from two configurations α^t and α^{t-1} as follows.

$$\forall \alpha^t, \alpha^{t-1} \in \mathrm{Conf}(A),\ \forall \mathbf{x} \in \mathbb{Z}^k :$$
$$\tilde{F}(\alpha^t, \alpha^{t-1})(\mathbf{x}) = f(\alpha^t(\mathbf{x} + \mathbf{n}_1), \dots, \alpha^t(\mathbf{x} + \mathbf{n}_m)) \ominus \alpha^{t-1}(\mathbf{x})$$

Hence,

$$\alpha^{t+1}(\mathbf{x}) = f(\alpha^t(\mathbf{x} + \mathbf{n}_1), \dots, \alpha^t(\mathbf{x} + \mathbf{n}_m)) \ominus \alpha^{t-1}(\mathbf{x})$$

holds for all $\mathbf{x} \in \mathbb{Z}^k$. On the other hand, by the assumption on the operator $\ominus$, the following equation also holds, which means that the state of each cell at time $t - 1$ is determined by the states of the neighboring cells at t and the state of the center cell at $t + 1$ by the same local function f.

$$\alpha^{t-1}(\mathbf{x}) = f(\alpha^t(\mathbf{x} + \mathbf{n}_1), \dots, \alpha^t(\mathbf{x} + \mathbf{n}_m)) \ominus \alpha^{t+1}(\mathbf{x})$$

In this sense, a second-order CA defined above is reversible for *any* local function $f : Q^m \to Q$ provided that the operator $\ominus$ is appropriately given. Of course, it is not in a traditional framework of CAs.

1.2   Objectives of This Book

The main objective of this book is to show that even quite simple reversible PCAs have rich capabilities of pattern generation and computing. The book presents attractive puzzles to readers on the problems of how fantastic and complex phenomena emerge from simple reversible microscopic laws. The problems range from finding various patterns having interesting features, to designing and creating reversible computers in reversible PCAs.

An important problem that is particularly focused on in this book is how we can compose reversible computers from very (or even extremely) simple reversible microscopic laws. Designing and creating reversible computers in a simple reversible PCA is not easy. Therefore, we have to find a pathway from a reversible microscopic law to reversible computers. To do so, it is convenient to set several implementation levels between the microscopic level and the macroscopic one as shown in Fig. 1.11. By this, the problem is decomposed into several simpler subproblems.

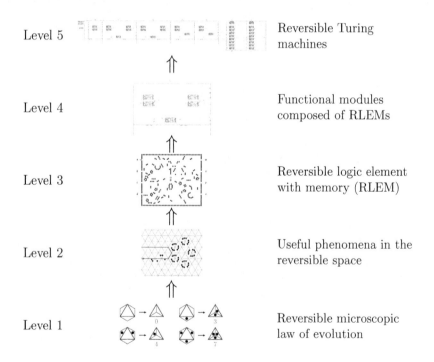

Fig. 1.11   Pathway from a simple reversible microscopic law to reversible computers.

In the bottom level, *i.e.*, Level 1, there are simple local transition rules of a PCA, which corresponds to a reversible microscopic physical law. In Level 2, various fundamental phenomena that emerge from the reversible microscopic law are observed. In Level 3, we implement suitable reversible logic elements using the observed phenomena. In Level 4, combining the reversible logic elements, functional modules for reversible computers are composed. In the top level, *i.e.*, Level 5, a reversible computing machine is systematically constructed by assembling the reversible functional modules. Whether we can successfully find a pathway from a reversible microscopic law to reversible computers depends on the choice of a system or an object in each level.

In Level 1, a local function of a reversible PCA is chosen. Of course, it is not known in advance which reversible PCA has sufficient capability for composing reversible computers. Therefore, we shall examine various reversible PCAs, and select seemingly good ones that give useful phenomena in the next level.

In Level 2, we try to find useful patterns and phenomena. First, periodic patterns and space-moving patterns of short periods are sought. Then, we make various experiments of interacting periodic and space-moving patterns. If useful phenomena are found, then, for example, a space-moving pattern is used as a signal, and periodic pattern are used to control signals.

In Level 3, we implement reversible logic elements. In this book, we implement a reversible logic element with memory (RLEM) rather than a reversible logic gate. We shall see that RLEMs can be implemented using a small number of useful phenomena found in Level 2. In addition, construction of reversible computing machines in the upper levels is greatly simplified than to use a reversible logic gate.

In Levels 4 and 5, reversible computers, such as reversible Turing machines (RTMs), are constructed. To do so, in Level 4, functional modules are composed out of RLEMs. Then, in Level 5, reversible computers are systematically constructed by assembling the modules.

In such a way, reversible computers will be constructed in several simple reversible PCAs systematically. When doing so, many sub-problems should be solved. Of course, there are possibilities of choosing other systems or objects in each level, By this, resulting reversible computers may become simpler or more efficient. Thus, one solution for composing complex systems in a simple reversible PCA brings new puzzling problems to readers. Presenting unsolved problems and future research problems, in which readers may be interested, is also one of the objectives of the book.

1.3   Using a General Purpose CA Simulator Golly

If one has created a configuration of a CA, he/she will surely want to know how it evolves. It can be done by hand, if the configuration is very small, and the number of steps is ten or so. However, tracing an evolving process by hand is a tedious task. Hence, a good CA simulator is necessary if one wants to observe evolution processes of large configurations for many steps.

We use a free CA simulator *Golly* [1] (Fig. 1.12) to investigate our reversible PCAs. The name "Golly" comes from GoL (the Game of Life), but a large variety of CAs can be simulated on it by giving a description of a local function of a CA. It is indeed an extraordinarily powerful simulator. It can deal with a huge configuration consisting of, say, millions of non-blank cells. In addition, it can simulate an evolving process for millions (or even billions) of steps with a very high speed, unless the simulated configuration contains a large chaotic (*i.e.*, random-like) pattern.

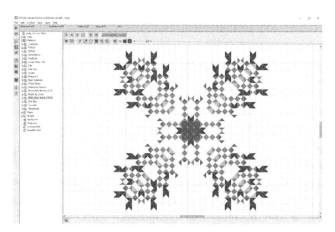

Fig. 1.12   Golly [1]. In this figure, a reversible elementary square PCA is simulated.

The PCAs and the configurations given in this book are implemented on Golly utilizing its great simulating capability. By this, correctness of their designs has been verified. In fact, without Golly, the study in this book was not possible. Readers are strongly recommended to use Golly, since evolution processes of configurations are easily viewed and understood by it. A zip file "Reversible_World_of_CAs.zip," which contains the files of the local functions of the proposed PCAs (called *rule files* in Golly), and the files of the configurations (called *pattern files*), is found in [21]. Readers can play with these files by downloading the Golly simulator file from [1], installing it on your computer, putting the zip file in the "patterns" folder of Golly, and accessing the pattern files in the zip file from it. Using Golly and solving exercise problems, readers will have an experience of composing reversible computers out of very simple microscopic phenomena. Readers can also design new configurations by themselves, and create unique functional objects on Golly.

Although Golly is a simulator for CAs, it can also be used to visualize computing processes of reversible Turing machines (RTMs) (Sec. 4.1), reversible counter machines (RCMs) (Sec. 4.2), and circuits composed of reversible logic elements with memory (RLEMs) (Sec. 4.3). In particular, the simulator for RLEM-circuits is useful for viewing computing processes of a large circuit for many steps. Files of these simulators and their pattern files are also contained in the zip file. They will be helpful for readers to understand behavior of RTMs, RCMs and RLEM circuits.

1.4   How to Read This Book

This book consists of nine chapters. Chapter 1 is an introduction. There, CAs as a model of a digital world are briefly described, and objectives of our study and how to read this book are explained.

Chapters 2–4 form the Part 1 of the book, where theoretical basis is given. Chapter 2 gives definitions on partitioned cellular automata (PCAs), in particular, a very simple subclass of them called elementary PCAs, and their reversible versions. Chapter 3 studies time-reversal symmetry (T-symmetry) in reversible elementary PCAs. It is a characteristic property that often appears in reversible systems, where the same evolution law holds for both forward and backward time directions. By this, an inverse functional module can be easily designed. Chapter 4 describes three models of reversible computing systems other than CAs. They are reversible Turing machines (RTMs), reversible counter machines (RCMs), and reversible logic elements with memory (RLEMs). When showing computational universality of reversible PCAs in the later chapters, these models will be used.

Chapters 5–9 form the Part 2 of the book, in which we can see that even very simple reversible PCAs have rich capabilities of pattern generation and computing. Chapter 5 shows fantastic phenomena generated by elementary PCAs. Chapter 6 investigates computational universality of four reversible elementary square PCAs having 16 states. In each of these PCAs, an RLEM is implemented using only a few kinds of small patterns and their interactions. Then an RTM is constructed out of the RLEM. Chapter 7 shows computational universality of four kinds of reversible elementary triangular PCAs having eight states. Despite the extreme simplicity of their local functions, RTMs are constructed using the composed RLEMs in these triangular PCAs. Chapter 8 studies an 81-state reversible square PCA. Though it has a larger number of states, any RCM can be realized in its cellular space concisely as a finite configuration. Chapter 9 discusses open problems and future research problems on reversible CAs, which will further clarify their capabilities. In addition to the problems given there, readers are encouraged to find new problems, and solve them.

Note that the first part (Chaps. 2–4) contains the theoretical basis that is necessary to derive useful results in the later chapters in a precise way. However, readers are recommended to read the second part first, since this part is more important. Fortunately, it is easy to understand the framework of PCAs. There are only two things that readers should know. The first one is the fact that their cellular spaces are as in Fig. 1.8. The second one is that their local transition rules are of the form shown in Fig. 1.9, which are applied to all the cells simultaneously. By them, almost all the results given in Chap. 5 are understandable. Using Golly discussed in Sec. 1.3 will be helpful to recognize the given fantastic phenomena.

In Chaps. 6 and 7, some background knowledge on RLEMs and RTMs are required. As for RLEMs, it is sufficient to know firstly that their operations are expressed by figures like Figs. 4.12 and 4.13 (or by Figs. 4.7 and 4.8 in the case of a rotary element (RE)). As for RTMs, it is sufficient to know their structure by Fig. 4.1, and the fact that RTMs are implemented by RLEMs as in Fig. 4.28 or 4.34. By them, readers can convince that RTMs are realized in the given reversible PCAs. Again, by using Golly, computing processes of the RTMs embedded in these PCAs can be viewed.

Likewise, in Chap. 8, some knowledge on RCMs are required. Also in this case, there are only two things that readers should know. They are the fact that the structure of RCMs is depicted as in Fig. 4.2, and the fact that their basic operations are simple ones as described in Sec. 4.2.3.1.

As noted above, after understanding an outline of Chaps. 5–8, readers can go to Chaps. 2–4 to know theoretical foundations. In addition, there is yet another good method. First, view evolving processes of the pattern files given in [21] using Golly without reading the book. By this, a rough outline of the book can be grasped. Then, read the book. Moreover, one can read a chapter or a section randomly. The book is organized in a logically consistent way if Chaps. 1–9 are read in this order, but there is no serious difficulty even if any chapter or section is read first.

1.5 Terminology and Notations

We give basic terminology and notations on logic, mathematics, and formal languages used in this book.

First, we explain notations on logic. Here, P, P_1 and P_2 are propositions, x is a variable, and $P(x)$ is a predicate with a free variable x.

$\neg P$	Negation of P
$P_1 \vee P_2$	Disjunction (logical OR) of P_1 and P_2
$P_1 \wedge P_2$	Conjunction (logical AND) of P_1 and P_2
$P_1 \Rightarrow P_2$	P_1 implies P_2
$P_1 \Leftrightarrow P_2$	P_1 if and only if P_2
$\forall x\,(P(x))$	For all x, $P(x)$ holds
$\exists x\,(P(x))$	There exists x such that $P(x)$ holds

When describing logical functions of logic gates and circuits, operations of NOT, OR and AND are expressed by $\bar{a}$, $a + b$, and $a \cdot b$, respectively, instead of $\neg$, $\vee$, and $\wedge$. Exclusive OR (XOR) is denoted by $a \oplus b$. Here, a and b are Boolean variables with a value 0 (false) or 1 (true).

Notations and symbols on set theory are as follows, where S, S_1 and S_2 are sets, a is an object, x is a variable, and $P(x)$ is a predicate with a free variable x.

$\emptyset$	The empty set
$a \in S$	a is an element of S
$S_1 \subseteq S_2$	S_1 is a subset (not necessarily a proper subset) of S_2
$S_1 \subset S_2$	S_1 is a proper subset of S_2
$S_1 \cup S_2$	The union of S_1 and S_2
$S_1 \cap S_2$	The intersection of S_1 and S_2
$S_1 - S_2$	The difference of S_1 and S_2
$S_1 \times S_2$	The Cartesian product of S_1 and S_2 ($S \times S$ is denoted by S^2)
2^S	The power set of S (*i.e.*, the set of all subsets of S)
$|S|$	The number of elements in S
$\{x \mid P(x)\}$	The set of all elements x that satisfy $P(x)$
$\mathbb{N}$	The set of all natural numbers (including 0)
$\mathbb{Z}$	The set of all integers

Terminology on relations and functions (mappings) is given below. Let S_1 and S_2 be sets. If $R \subseteq S_1 \times S_2$, then R is called a *(binary) relation*. If $(x, y) \in R$, it is also written by xRy. Generally, let $S_1, \ldots, S_n$ be sets, and if $R \subseteq S_1 \times \cdots \times S_n$, then R is called an *n-ary relation*. For the case $R \subseteq S^2$, we define $R^{(n)}$ ($n \in \mathbb{N}$) recursively as follows: $R^{(0)} = \{(x,x) \mid x \in S\}$, and $R^{(i+1)} = \{(x,y) \mid \exists z \in S\ ((x,z) \in R\ \land\ (z,y) \in R^{(i)})\}$ ($i \in \mathbb{N}$). Then, R^* and R^+ are defined below.

R^* The *reflexive and transitive closure* of the relation R, *i.e.*, $R^* = \bigcup_{i=0}^{\infty} R^{(i)}$

R^+ The *transitive closure* of the relation R, *i.e.*, $R^+ = \bigcup_{i=1}^{\infty} R^{(i)}$

A relation $f \subseteq S_1 \times S_2$ is called a *partial function* (or *partial mapping*) from S_1 to S_2, if it satisfies

$$\forall x \in S_1\ \forall y_1, y_2 \in S_2\ (((x,y_1) \in f) \land ((x,y_2) \in f) \Rightarrow (y_1 = y_2)),$$

which means that for each $x \in S_1$ there exists at most one $y \in S_2$ such that $(x,y) \in f$. A partial function f from S_1 to S_2 is denoted by $f : S_1 \to S_2$, where the sets S_1 and S_2 are called the *domain* and the *codomain* of f, respectively. As usual, $(x,y) \in f$ is denoted by $y = f(x)$. If $(x,y) \notin f$ for all $y \in S_2$, then $f(x)$ is undefined. The notation $x \mapsto f(x)$ indicates x maps to $f(x)$.

A partial function f is called a *(total) function* (or *(total) mapping*) from S_1 to S_2, if it further satisfies

$$\forall x \in S_1\ \exists y \in S_2\ ((x,y) \in f)$$

Let $f : S_1 \to S_2$ and $g : S_2 \to S_3$ be total functions. The *composition* $g \circ f : S_1 \to S_3$ of the functions f and g is defined as follows.

$$\forall x \in S_1\ (g \circ f(x) = g(f(x)))$$

An *identity function* over S is a total function $f_\mathrm{I} : S \to S$ that satisfies

$$\forall x \in S\ (f_\mathrm{I}(x) = x).$$

For a total function $f : S \to S$, we define f^n ($n \in \mathbb{N}$) recursively as follows.

$$\forall x \in S\ (f^0(x) = x)$$
$$\forall x \in S\ (f^{n+1}(x) = f(f^n(x)))\ \ (n \in \mathbb{N})$$

Let f and g be total functions such that $f : A_1 \to B$, $g : A_2 \to B$, and $A_1 \subseteq A_2$. If $\forall x \in A_1(g(x) = f(x))$ holds, then g is called an *extension* of f, and f is called a *restriction* of g to A_1, which is denoted by $g|_{A_1}$.

A partial function $f : S_1 \to S_2$ is called *injective* if

$$\forall x_1, x_2 \in S_1\ \forall y \in S_2\ ((f(x_1) = y) \land (f(x_2) = y) \Rightarrow (x_1 = x_2))$$

A partial function $f : S_1 \to S_2$ is called *surjective* if

$$\forall y \in S_2\ \exists x \in S_1\ (f(x) = y)$$

A partial function $f : S_1 \rightarrow S_2$ that is both injective and surjective is called *bijective*. If a total function $f : S_1 \rightarrow S_2$ is injective (surjective, or bijective, respectively), then it is called an *injection* (*surjection*, or *bijection*).

Let $f : S_1 \rightarrow S_2$ be an injection. The *inverse partial function* of f is denoted by $f^{-1} : S_2 \rightarrow S_1$, and is defined as follows.

$$\forall x \in S_1 \ \forall y \in S_2 \ (f(x) = y \ \Leftrightarrow \ f^{-1}(y) = x)$$

Hence, $f^{-1}(f(x)) = x$ holds for all $x \in S_1$, and f^{-1} is an injective partial function. Note that, for $y_0 \in S_2$, if there is no $x \in S_1$ such that $f(x) = y_0$, then $f^{-1}(y_0)$ is undefined. If f is a bijection, then f^{-1} is totally defined, and thus called the *inverse function* of f, which is also a bijection.

Notations on formal languages are given below. A nonempty finite set of symbols is called an *alphabet*. Let Σ be an alphabet. A finite sequence of symbols $a_1 \cdots a_n$ ($n \in \mathbb{N}$) taken from Σ is called a *string* (or *word*) over the alphabet Σ. The *concatenation of strings* w_1 and w_2 is denoted by $w_1 \cdot w_2$ (usually $\cdot$ is omitted). The length of a string w is denoted by $|w|$. Hence, if $w = a_1 \cdots a_n$, then $|w| = n$. We denote the *empty string* (*i.e.*, the string of length 0) by λ. For a symbol a, we use a^n to denote the string consisting of n repetitions of a ($n \in \mathbb{N}$). We define the set of strings Σ^n ($n \in \mathbb{N}$) recursively as follows: $\Sigma^0 = \{\lambda\}$, and $\Sigma^{i+1} = \{aw \mid a \in \Sigma \ \wedge \ w \in \Sigma^i\}$ ($i \in \mathbb{N}$). Then, Σ^* and Σ^+ are defined below.

Σ^* The set of all strings over Σ including λ, *i.e.*, $\Sigma^* = \bigcup_{i=0}^{\infty} \Sigma^i$

Σ^+ The set of all strings over Σ of positive length, *i.e.*, $\Sigma^+ = \bigcup_{i=1}^{\infty} \Sigma^i$

A subset of Σ^* is called a *(formal) language* over the alphabet Σ.

Let L, L_1 and L_2 be languages over Σ. The *concatenation of languages* of L_1 and L_2 is defined by $L_1 \cdot L_2 = \{w_1 w_2 \mid w_1 \in L_1 \ \wedge \ w_2 \in L_2\}$ (usually $\cdot$ is omitted). We define the language L^n recursively in a similar manner as in Σ^n: $L^0 = \{\lambda\}$, and $L^{i+1} = L \cdot L^i$ ($i \in \mathbb{N}$). Then, L^* and L^+ are as follows: $L^* = \bigcup_{i=0}^{\infty} L^i$, and $L^+ = \bigcup_{i=1}^{\infty} L^i$.

Finally, we give two kinds of distance measures on the 2-dimensional plane. Let $(x_1, y_1), (x_2, y_2) \in \mathbb{Z}^2$ be two points. The *Manhattan distance* dist_M and the *Chebyshev distance* dist_C between these points are defined as follows.

$$\text{dist}_M((x_1, y_1), (x_2, y_2)) = |x_1 - x_2| + |y_1 - y_2|$$
$$\text{dist}_C((x_1, y_1), (x_2, y_2)) = \max\{|x_1 - x_2|, |y_1 - y_2|\}$$

When measuring the distance between two cells in a 2-dimensional CA with the von Neumann neighborhood, we use the Manhattan distance. In the case of the Moore neighborhood, it is appropriate to use the Chebyshev distance.

1.6   Exercises

Each chapter has exercise problems. Readers are encouraged to solve them to understand reversible CAs and related systems more deeply. There are two categories of exercises: "Paper-and-pencil exercises" and "Golly exercises." For the problems in the latter category, the CA simulator Golly [1] is needed (see Sec. 1.3) for dealing with more complex phenomena than the former category. Difficulty of each exercise problem is indicated by the number of *'s, *i.e.*, Easy: *, Intermediate: **, and Difficult: ***. Solutions for selected exercise problems are found in [21].

1.6.1   *Paper-and-pencil exercises*

Exercise 1.1.* Consider the pattern called a *toad* [4] in GoL shown in Fig. 1.13. Write its evolving process in GoL for $0 \le t \le 2$.

$t = 0$

Fig. 1.13   A pattern called a *toad* in GoL.

Exercise 1.2.** Consider the Fredkin's self-replicating CA (Example 1.1). Prove that "any" initial pattern replicates in this cellular space after some time steps. Also show the relation between the total number of replicated patterns and time.

1.6.2   *Golly exercises*

Exercise 1.3.* Consider the Fredkin's self-replicating CA. Observe that an arbitrarily chosen pattern replicates in this cellular space using Golly.

Exercise 1.4.* Consider the pattern shown in Fig. 1.14 in GoL, which is called an *R-pentomino*. Observe the evolving process of it using Golly.

$t = 0$

Fig. 1.14   A pattern called an *R-pentomino* in GoL.

PART 1
Theoretical Basis

Chapter 2

Elementary Partitioned Cellular Automata

A partitioned cellular automaton (PCA) is a special subclass of usual CAs where each cell is divided into several parts, and the cell changes its state depending on the states of specially designated parts of its neighbor cells. It was first proposed in [22] to show computational universality of 1-dimensional reversible cellular automaton. The reason why we use PCAs is that they make it very easy to design reversible CAs. In this chapter, we introduce a 2-dimensional square partitioned cellular automaton (SPCA), and a triangular partitioned cellular automaton (TPCA). We then define an elementary SPCA (ESPCA) and an elementary TPCA (ETPCA), which are simplest subclasses of SPCAs and TPCAs. Actually, they have extremely simple local transition functions. These elementary PCAs are the main concern of this book, since they show interesting behavior and have high capability of computing even if the reversibility constraint is added. These results will be shown in the later chapters. Here, we give formal definitions of them, show how they evolve, and investigate their basic properties.

2.1 Elementary Square Partitioned Cellular Automaton

First, we give definitions of a square partitioned cellular automaton (SPCA), and show a method of designing reversible SPCAs. We then introduce an elementary SPCA (ESPCA), the simplest subclass of SPCAs, and observe how configurations evolve in reversible ESPCAs.

2.1.1 *Square partitioned cellular automaton (SPCA)*

In a partitioned cellular automaton (PCA), each cell is divided into several parts, whose number is equal to the number of neighbor cells. A square cell of a 4-neighbor SPCA has thus four parts, and its cellular space is depicted as in Fig. 2.1(a) (but, in the following, partitioning (*i.e.*, thin) lines are often omitted). Each part has its own state set. Hence, the set of states of one cell is the Cartesian product of the sets of states of the four parts. In an SPCA, a cell changes its state depending on the top part of the south-neighbor cell, the right part of the west cell, the bottom part of the north cell, and the left part of the east cell as shown in Fig. 2.1(b). Note that, in the case of a 4-neighbor SPCA, the next state of a cell does not depend on the state of the cell itself. Thus, it is a special subcase of the von Neumann neighborhood. Each part of a cell can be interpreted as an "output port" to the corresponding neighbor cell. Hence, the state in the part is just a signal moving to the neighbor cell.

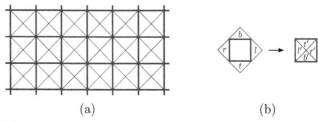

(a)   (b)

Fig. 2.1   (a) Cellular space of a 4-neighbor SPCA, and (b) a local transition rule that represents $f(t, r, b, l) = (t', r', b', l')$.

Definition 2.1. A *4-neighbor square partitioned cellular automaton* (SPCA) is a system defined by

$$P = (\mathbb{Z}^2, (T, R, B, L), ((0, -1), (-1, 0), (0, 1), (1, 0)), f, (\#_1, \#_2, \#_3, \#_4))$$

Here, $\mathbb{Z}^2$ is the set of all points with integer coordinates where cells are placed. The items T, R, B, and L are non-empty finite sets of states of the top, right, bottom, and left parts of a cell. The set of states of a cell is thus $Q = T \times R \times B \times L$. The quadruple $((0, -1), (-1, 0), (0, 1), (1, 0))$ is a *neighborhood* of each cell, and $f : Q \to Q$ is a *local function*. The state $(\#_1, \#_2, \#_3, \#_4) \in Q$ is a *quiescent state* that satisfies $f(\#_1, \#_2, \#_3, \#_4) = (\#_1, \#_2, \#_3, \#_4)$. It is also called a *blank state*. We also allow an SPCA where no quiescent state exists, or no quiescent state is specified.

The local function determines the transition of each cell depending on the states of adjacent parts of its neighboring cells. If $f(t, r, b, l) = (t', r', b', l')$ holds for $(t, r, b, l), (t', r', b', l') \in Q$, this relation is called a *local transition rule* of P. It is also indicated as in Fig. 2.1(b). The local function f is thus defined by a set of local transition rules.

Definition 2.2. Let $P = (\mathbb{Z}^2, (T, R, B, L), ((0, -1), (-1, 0), (0, 1), (1, 0)),$ $f)$ be an SPCA. A *configuration* of P is a function $\alpha : \mathbb{Z}^2 \to Q$. The set of all configurations of P is denoted by $\mathrm{Conf}(P) = \{\alpha \,|\, \alpha : \mathbb{Z}^2 \to Q\}$.

Since a configuration α is a function $\mathbb{Z}^2 \to Q$, the value $\alpha(\mathbf{x})$ gives the state of the cell at $\mathbf{x} \in \mathbb{Z}^2$.

Definition 2.3. Let $P = (\mathbb{Z}^2, (T, R, B, L), ((0, -1), (-1, 0), (0, 1), (1, 0)),$ $f, (\#_1, \#_2, \#_3, \#_4))$ be an SPCA, and $\alpha \in \mathrm{Conf}(P)$. The set $\mathrm{supp}(\alpha) = \{\mathbf{x} \,|\, \alpha(\mathbf{x}) \neq (\#_1, \#_2, \#_3, \#_4)\}$ is called the *support* of α. We say α is a *finite configuration* if $\mathrm{supp}(\alpha)$ is finite. Otherwise, it is an *infinite configuration*.

Definition 2.4. Let P be an SPCA with a quiescent state, and α be a finite configuration of P. If $\mathrm{supp}(\alpha) \neq \emptyset$, we define $\mathrm{width}(\alpha)$ and $\mathrm{height}(\alpha)$ as follows.

$$\mathrm{width}(\alpha) = \max\{x_1 - x_2 + 1 \mid \exists y_1, y_2((x_1, y_1), (x_2, y_2) \in \mathrm{supp}(\alpha))\}$$
$$\mathrm{height}(\alpha) = \max\{y_1 - y_2 + 1 \mid \exists x_1, x_2((x_1, y_1), (x_2, y_2) \in \mathrm{supp}(\alpha))\}$$

The *diameter* of α is defined as follows.

$$\mathrm{diam}(\alpha) = \max\{\mathrm{width}(\alpha), \mathrm{height}(\alpha)\}$$

When $\mathrm{supp}(\alpha) = \emptyset$, α is called a *null configuration*, and is denoted by α_{null}. We define $\mathrm{diam}(\alpha_{\mathrm{null}}) = \mathrm{width}(\alpha_{\mathrm{null}}) = \mathrm{height}(\alpha_{\mathrm{null}}) = 0$.

Next, we define a global function F of an SPCA. It is the mapping that derives the next configuration from the present configuration. Therefore, the state of an entire cellular space evolves step by step by applying F repeatedly. The global function is induced by a local function as given in the following definition.

Definition 2.5. Let $P = (\mathbb{Z}^2, (T, R, B, L), ((0, -1), (-1, 0), (0, 1), (1, 0)),$ $f)$ be an SPCA. Let $\mathrm{pr}_T : Q \to T$ be the *projection function* that satisfies $\mathrm{pr}_T(t, r, b, l) = t$ for all $(t, r, b, l) \in Q$. The projection functions $\mathrm{pr}_R : Q \to R$, $\mathrm{pr}_B : Q \to B$ and $\mathrm{pr}_L : Q \to L$ are defined similarly. The *global function* $F : \mathrm{Conf}(P) \to \mathrm{Conf}(P)$ of P is defined as the one that satisfies the following.

$$\forall \alpha \in \mathrm{Conf}(P), \forall \mathbf{x} \in \mathbb{Z}^2 :$$
$$F(\alpha)(\mathbf{x}) = f(\mathrm{pr}_T(\alpha(\mathbf{x} + (0, -1))), \mathrm{pr}_R(\alpha(\mathbf{x} + (-1, 0))),$$
$$\mathrm{pr}_B(\alpha(\mathbf{x} + (0, 1))), \mathrm{pr}_L(\alpha(\mathbf{x} + (1, 0))))$$

The *evolution process* starting from α is a sequence of configurations $\alpha, F(\alpha), F^2(\alpha), \ldots$.

Remark 2.1. A point $(x, y) \in \mathbb{Z}^2$ is called an *even parity point* (*odd parity point*, respectively) if $x + y$ is even (odd). A cell at an even parity point (odd parity point, respectively) is called an *even parity cell* (odd parity cell). From the definition of the global function F of an SPCA, it is clear that the states of even parity cells (odd parity cells, respectively) at $t + 1$ are determined depending only on the states of odd parity cells (even parity cells) at t. It means that the states of even parity cells at $t = 0$, and those of odd parity cells at $t = 0$ will never interact, *i.e.*, these two sets evolve independently. Therefore, if we want to design an initial configuration that performs some operation, it is sufficient to use only even parity (or odd parity) cells. However, if we do so, no *stable* configuration, which does not change its configuration, can exist. Since a stable pattern, which is a non-blank segment of a stable configuration, is often convenient to design the initial configuration, we allow for using them in the following.

2.1.2   *Reversible SPCA*

We define the notion of reversibility as follows. Here, a reversible SPCA is simply defined by an injective SPCA. Note that a detailed discussion on the definition of a *reversible CA* is found in Sec. 10.3 of [9].

Definition 2.6. Let P be an SPCA whose global function is F. If F is an injection, then P is called a *reversible SPCA*.

The next theorem shows that, in an SPCA, injectivity of the global function is equivalent to injectivity of the local function. It was first proved for a 3-neighbor 1-dimensional PCA in [22]. A proof for a PCA of any dimension and of any neighborhood is given in [9]. Here, we write a proof for a 4-neighbor SPCA.

Theorem 2.1. *Let P be a 4-neighbor SPCA. Its global function F is injective if and only if its local function f is injective.*

Proof. First, we show if F is injective, then f is injective. Assume, on the contrary, f is not injective. Then, there are $(t_1, r_1, b_1, l_1), (t_2, b_2, b_2, l_2) \in Q = T \times R \times B \times L$ such that $f(t_1, r_1, b_1, l_1) = f(t_2, b_2, b_2, l_2)$ and $(t_1, r_1, b_1, l_1) \neq (t_2, b_2, b_2, l_2)$ hold. Let $(\hat{t}, \hat{r}, \hat{b}, \hat{l}) \in Q$ be an arbitrarily chosen element. Define two configurations α_i ($i \in \{1, 2\}$) as follows, where

$\mathbf{x} \in \mathbb{Z}^2$.

$$\mathrm{pr}_T(\alpha_i(\mathbf{x})) = \begin{cases} t_i & \text{if } \mathbf{x} = (0, -1) \\ \hat{t} & \text{elsewhere} \end{cases} \qquad \mathrm{pr}_R(\alpha_i(\mathbf{x})) = \begin{cases} r_i & \text{if } \mathbf{x} = (-1, 0) \\ \hat{r} & \text{elsewhere} \end{cases}$$

$$\mathrm{pr}_B(\alpha_i(\mathbf{x})) = \begin{cases} b_i & \text{if } \mathbf{x} = (0, 1) \\ \hat{b} & \text{elsewhere} \end{cases} \qquad \mathrm{pr}_L(\alpha_i(\mathbf{x})) = \begin{cases} l_i & \text{if } \mathbf{x} = (1, 0) \\ \hat{l} & \text{elsewhere} \end{cases}$$

Since $(t_1, r_1, b_1, l_1) \neq (t_2, b_2, b_2, l_2)$, $\alpha_1 \neq \alpha_2$ holds. By the definition of F, the following holds for $i \in \{1, 2\}$ and for all $\mathbf{x} \in \mathbb{Z}^2$.

$$F(\alpha_i)(\mathbf{x}) = \begin{cases} f(t_i, r_i, b_i, l_i) & \text{if } \mathbf{x} = (0, 0) \\ f(\hat{t}, \hat{r}, \hat{b}, \hat{l}) & \text{elsewhere} \end{cases}$$

Therefore $F(\alpha_1) = F(\alpha_2)$, but it contradicts the assumption F is injective.

Second, we show if f is injective, then F is injective. Assume F is not injective. Therefore, there are configurations α_1 and α_2 such that $F(\alpha_1) = F(\alpha_2)$ and $\alpha_1 \neq \alpha_2$. Let $n_T = (0, -1), n_R = (-1, 0), n_B = (0, 1)$, and $n_L = (1, 0)$. Since $\alpha_1 \neq \alpha_2$, there exist $\mathbf{x}_0 \in \mathbb{Z}^2$ and $X \in \{T, R, B, L\}$ that satisfy $\mathrm{pr}_X(\alpha_1(\mathbf{x}_0 + n_X)) \neq \mathrm{pr}_X(\alpha_2(\mathbf{x}_0 + n_X))$. From the definition of F, the following holds ($i \in \{1, 2\}$).

$$F(\alpha_i)(\mathbf{x}_0) = f(\mathrm{pr}_T(\alpha_i(\mathbf{x}_0 + n_T)), \mathrm{pr}_R(\alpha_i(\mathbf{x}_0 + n_R)),$$
$$\mathrm{pr}_B(\alpha_i(\mathbf{x}_0 + n_B)), \mathrm{pr}_L(\alpha_i(\mathbf{x}_0 + n_L)))$$

Since f is injective and $\mathrm{pr}_X(\alpha_1(\mathbf{x}_0 + n_X)) \neq \mathrm{pr}_X(\alpha_2(\mathbf{x}_0 + n_X))$ for some $X \in \{T, R, B, L\}$, $F(\alpha_1)(\mathbf{x}_0) \neq F(\alpha_2)(\mathbf{x}_0)$ must hold. But, it contradicts the assumption $F(\alpha_1) = F(\alpha_2)$. $\qquad\square$

It is easy to show the next theorem [9, 22], which says that an SPCA is a subclass of a traditional CA (Definition 1.1) with square cells.

Theorem 2.2. *Let $P = (\mathbb{Z}^2, (T, R, B, L), ((0, -1), (-1, 0), (0, 1), (1, 0)), f)$ be a 4-neighbor SPCA, and F be its global function. Then, there is a CA $A = (\mathbb{Z}^2, (T \times R \times B \times L), ((0, -1), (-1, 0), (0, 1), (1, 0))), \hat{f})$ such that $\hat{F} = F$, where $\hat{F}$ is the global function of A.*

Proof. Let $Q = (T \times R \times B \times L)$, and let pr_T, pr_R, pr_B, and pr_L be projection functions given in Definition 2.5. Define $\hat{f}$ as follows.

$$\forall q_1, q_2, q_3, q_4 \in Q : \hat{f}(q_1, q_2, q_3, q_4) = f(\mathrm{pr}_T(q_1), \mathrm{pr}_R(q_2), \mathrm{pr}_B(q_3), \mathrm{pr}_L(q_4))$$

Then, for any $\alpha \in \mathrm{Conf}(P)$, and $\mathbf{x} \in \mathbb{Z}^2$,

$$\hat{F}(\alpha)(\mathbf{x}) = \hat{f}(\alpha(\mathbf{x} + (0, -1)), \alpha(\mathbf{x} + (-1, 0)), \alpha(\mathbf{x} + (0, 1)), \alpha(\mathbf{x} + (1, 0)))$$
$$= f(\mathrm{pr}_T(\alpha(\mathbf{x} + (0, -1))), \mathrm{pr}_R(\alpha(\mathbf{x} + (-1, 0))),$$
$$\mathrm{pr}_B(\alpha(\mathbf{x} + (0, 1))), \mathrm{pr}_L(\alpha(\mathbf{x} + (1, 0))))$$
$$= F(\alpha)(\mathbf{x})$$

holds. Hence, $\hat{F} = F$. $\qquad\square$

By the above two theorems, we can easily obtain a reversible CA by designing a PCA whose local function is injective. Note that injectivity of a local function is tested by checking if there is no pair of distinct local transition rules whose right-hand sides are the same.

2.1.3 *Elementary SPCA (ESPCA)*

We define a subclass of an SPCA that satisfies the following conditions: Its local function is rotation-symmetric, and each of the four parts of a cell has two states. It is called an elementary SPCA (ESPCA) as in the case of a 1-dimensional elementary cellular automaton (ECA) [23]. First, the notion of rotation-symmetry is defined.

Definition 2.7. Let $P = (\mathbb{Z}^2, (T, R, B, L), ((0, -1), (-1, 0), (0, 1), (1, 0)),$ $f)$ be an SPCA. The SPCA P is called *rotation-symmetric* if the following conditions (1) and (2) hold.

(1) $T = R = B = L$
(2) $\forall (t, r, b, l), (t', r', b', l') \in T \times R \times B \times L :$
     $f(t, r, b, l) = (t', r', b', l') \Rightarrow f(r, b, l, t) = (r', b', l', t')$

It means that for each local transition rule shown in Fig. 2.1(b), there are rules obtained by rotating both sides of it by 90, 180 and 270 degrees.

Definition 2.8. Let $P = (\mathbb{Z}^2, (T, R, B, L), ((0, -1), (-1, 0), (0, 1), (1, 0)),$ $f)$ be an SPCA. We say P is an *elementary square partitioned cellular automaton* (ESPCA), if $T = R = B = L = \{0, 1\}$, and it is rotation-symmetric. Hence, it is a 16-state SPCA. The set of all *configurations* of an ESPCA is denoted by $\mathrm{Conf}_{\mathrm{E4}} = \{\alpha \mid \alpha : \mathbb{Z}^2 \to \{0, 1\}^4\}$.

Since an ESPCA is rotation-symmetric, its local function $f : \{0, 1\}^4 \to \{0, 1\}^4$ is defined by only six local transition rules. They are described by giving the following six values.

$f(0, 0, 0, 0), \; f(0, 0, 1, 0), \; f(0, 0, 1, 1), \; f(1, 0, 1, 0), \; f(0, 1, 1, 1), \; f(1, 1, 1, 1)$

Here, $f(0, 0, 1, 0), f(0, 0, 1, 1), f(0, 1, 1, 1) \in \{0, 1\}^4$. On the other hand, $f(1, 0, 1, 0) \in \{(0, 0, 0, 0), (0, 1, 0, 1), (1, 0, 1, 0), (1, 1, 1, 1)\}$ and $f(0, 0, 0, 0),$ $f(1, 1, 1, 1) \in \{(0, 0, 0, 0), (1, 1, 1, 1)\}$, since it is rotation-symmetric. Hence, there are $16^3 \times 4 \times 2^2 = 65,536$ ESPCAs in total.

Reading the 4-bit values of $f(0, 0, 0, 0), \quad f(0, 0, 1, 0), \quad f(0, 0, 1, 1),$ $f(1, 0, 1, 0), \quad f(0, 1, 1, 1), \quad f(1, 1, 1, 1)$ as six binary numbers, we express an ESPCA by a 6-digit hexadecimal *identification number uvwxyz* as in

Fig. 2.2. For example, if $f(0, 0, 1, 0) = (1, 0, 0, 1)$, then $v = 9$. An ESPCA with the identification number $uvwxyz$ is denoted by ESPCA-$uvwxyz$. Its local and global functions are also denoted by f_{uvwxyz} and F_{uvwxyz}, respectively, using the identification number. Note that, in the following, states 0 and 1 are represented by a blank and • (*particle*), respectively.

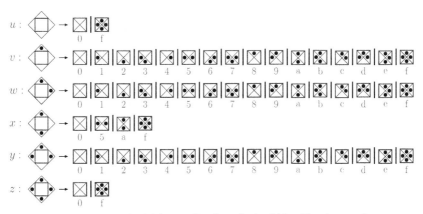

Fig. 2.2   Expressing an ESPCA by a 6-digit hexadecimal identification number $uvwxyz$. States 0 and 1 are represented by a blank and •. Vertical bars indicate alternatives of the right-hand side of each local transition rule.

Example 2.1. The six local transition rules shown in Fig. 2.3 define the local function f_{02c5bf} of ESPCA-02c5bf. It is easy to see that f_{02c5bf} is injective. Therefore, ESPCA-02c5bf is reversible.

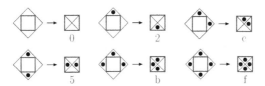

Fig. 2.3   Six local transition rules that define the local function of ESPCA-02c5bf.

We now introduce another important notion besides reversibility. It is "conservativeness" of an ESPCA. It is an analog of various conservation laws in physics such as conservation of mass, energy, momentum, *etc.*

Definition 2.9. An ESPCA P is called *conservative* if the number of state 1's (*i.e.*, particles) is conserved in each of its local transition rules.

For example, ESPCA-02c5bf in Example 2.1 is conservative. It is easy to see the following theorem [24] (its proof is omitted).

Theorem 2.3. *Let P be ESPCA-uvwxyz.*

(1) *P is reversible if and only if the following holds.*
   $(u, z) \in \{(0, f), (f, 0)\} \wedge x \in \{5, a\} \wedge (v, w, y) \in (A \times B \times C \cup A \times C \times B \cup B \times A \times C \cup B \times C \times A \cup C \times A \times B \cup C \times B \times A),$ *where* $A = \{1, 2, 4, 8\}, B = \{3, 6, 9, c\}, C = \{7, b, d, e\}$

(2) *P is conservative if and only if the following holds.*
   $u = 0 \wedge v \in \{1, 2, 4, 8\} \wedge w \in \{3, 5, 6, 9, a, c\} \wedge x \in \{5, a\} \wedge y \in \{7, b, d, e\} \wedge z = f$

(3) *P is reversible and conservative if and only if the following holds.*
   $u = 0 \wedge v \in \{1, 2, 4, 8\} \wedge w \in \{3, 6, 9, c\} \wedge x \in \{5, a\} \wedge y \in \{7, b, d, e\} \wedge z = f$

By above, we observe the total numbers of reversible, conservative, and reversible and conservative ESPCAs are 1536, 192, and 128, respectively. For example, ESPCA-02c5bf in Example 2.1 is reversible and conservative.

Next, we define the notion of population for finite configurations of ESPCAs. It gives the total number of particles contained in a configuration.

Definition 2.10. The *population* of a cell in the state $(t, r, b, l) \in \{0, 1\}^4$ is defined by $\mathrm{popul}_4(t, r, b, l) = t + r + b + l$. Let $\alpha \in \mathrm{Conf}_{E4}$ be a finite configuration of an ESPCA. Then, the *population* of α is defined as follows.

$$\mathrm{popul}(\alpha) = \sum_{\mathbf{x} \in \mathrm{supp}(\alpha)} \mathrm{popul}_4(\alpha(\mathbf{x}))$$

In a conservative ESPCA, the population of a finite configuration is kept constant throughout its evolution process.

2.1.4   *Evolution of configurations in ESPCA*

Here, we consider ESPCAs that have a quiescent state $(0, 0, 0, 0)$, and investigate how finite configurations in irreversible and reversible ESPCAs evolve. As we shall see, finite configurations can be classified into a few categories depending on their evolving processes.

Definition 2.11. Let P be an ESPCA whose global function is F. A finite configuration α is called *periodic* in P, if there exists an integer $m > 0$ that satisfies $F^m(\alpha) = \alpha$. The minimum of such m is the *period* of α. If there exist integers $m > n \geq 0$ that satisfy $F^m(\alpha) = F^n(\alpha)$, then α is called *pre-periodic*. By above, if α is periodic, it is pre-periodic. We say that α is *properly pre-periodic*, if it is pre-periodic, but not periodic.

Definition 2.12. Let P be an ESPCA whose global function is F. A finite configuration α is called *space-moving* in P, if there exist an integer $m > 0$ and $\mathbf{x}_0 \in \mathbb{Z}^2 - \{(0,0)\}$ such that $F^m(\alpha)(\mathbf{x} + \mathbf{x}_0) = \alpha(\mathbf{x})$ holds for all $\mathbf{x} \in \mathbb{Z}^2$. The minimum of such m is the *period* of α. If m is the period of α, $\mathbf{x}_0$ is called a *displacement vector* of α in one period. If there exist integers $m > n \geq 0$ and $\mathbf{x}_0 \in \mathbb{Z}^2 - \{(0,0)\}$ such that $F^m(\alpha)(\mathbf{x} + \mathbf{x}_0) = F^n(\alpha)(\mathbf{x})$, then α is called *pre-space-moving*. By above, if α is space-moving, it is pre-space-moving. We say that α is *properly pre-space-moving*, if it is pre-space-moving, but not space-moving. Note that among space-moving configurations, particularly useful one is named a *glider* following the terminology used in the GoL (Game of Life) [4, 5].

For a space-moving configuration, we can define its speed as the shifting distance divided by the period, where the distance is measured by the Manhattan distance (Sec. 1.5). It should be noted that the speed does not exceed unit distance per step, since each cell of SPCA changes its state depending only on its adjacent cells. This limit is called the *speed of light*, which is represented by c. Thus, in the following, the speed of a space-moving pattern is expressed in terms of c.

Definition 2.13. Let P be an ESPCA whose global function is F. Assume α is a space-moving configuration, *i.e.*, there exist an integer $m > 0$ and $\mathbf{x}_0 \in \mathbb{Z}^2 - \{(0,0)\}$ such that $F^m(\alpha)(\mathbf{x} + \mathbf{x}_0) = \alpha(\mathbf{x})$. The *speed* of α is defined below, where $\mathrm{dist}_{\mathrm{M}}(\mathbf{x}, \mathbf{y})$ is the Manhattan distance between $\mathbf{x}$ and $\mathbf{y}$.

$$c \cdot \mathrm{dist}_{\mathrm{M}}(\mathbf{x}_0, (0,0))/m$$

Definition 2.14. Let P be an ESPCA whose global function is F. A finite configuration α is called *diameter-growing* in P, if the following holds.

$$\forall d > 0 \; \exists t > 0 \; (\mathrm{diam}(F^t(\alpha)) > d)$$

Note that a diameter-growing configuration is one such that the diameters of configurations in its evolving process are unbounded, though the diameter may not increase monotonically.

Besides the notion of a diameter-growing configuration, there are notions of population-growing and population-bounded configurations.

Definition 2.15. Let P be an ESPCA whose global function is F. A finite configuration α in P is called a *population-growing* configuration, if the following holds.

$$\forall m > 0 \; \exists t > 0 \; (\mathrm{popul}(F^t(\alpha)) > m)$$

If the above does not hold, *i.e.*, if

$$\exists m > 0 \; \forall t > 0 \; (\text{popul}(F^t(\alpha)) \leq m)$$

then α is called a *population-bounded configuration*.

Apparently, a population-growing configuration is a diameter-growing one, but not *vice versa*.

We now show the following theorem for ESPCAs that are not necessarily reversible. It says that every configuration of an ESPCA belongs to one of the three categories.

Theorem 2.4. *Let P be an ESPCA, which is generally irreversible. If α is a finite configuration of P, then it is either pre-periodic, pre-space-moving, or diameter-growing.*

Proof. Let F be the global function of P. Consider the case where the following holds, and the case where it does not hold.

$$\forall d > 0 \; \exists t > 0 \; (\text{diam}(F^t(\alpha)) > d)$$

In the first case where the above holds, α is diameter-growing. Next, consider the second case where the negation of the above holds:

$$\exists d > 0 \; \forall t > 0 \; (\text{diam}(F^t(\alpha)) \leq d)$$

This formula means that there are infinitely many instances of t's that satisfy $\text{diam}(F^t(\alpha)) \leq d$. Therefore, there exists $t_1 > t_2 > 0$ such that $F^{t_1}(\alpha) = F^{t_2}(\alpha)$, or $F^{t_1}(\alpha)$ is a translation of $F^{t_2}(\alpha)$. This is because the total number of different configurations whose diameter is less than or equal to d is finite except translated ones. In the former sub-case α is pre-periodic, while in the latter sub-case α is pre-space-moving. $\square$

Example 2.2. We consider ESPCA-09458f, whose local function is given in Fig. 2.4. It has a quiescent state $(0, 0, 0, 0)$. It is an irreversible ESPCA, since the third local transition rule has the same right-hand side as the fifth one, if the latter is rotated by 90 degrees clockwise.

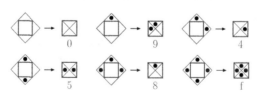

Fig. 2.4   Local function of irreversible ESPCA-09458f.

Configurations shown in Fig. 2.5 are all pre-periodic. Among them, the configurations at $t = 0$ and 1 are properly pre-periodic, while those at $t \geq 2$ are periodic ones whose period is 4. Figure 2.6 shows pre-space-moving configurations. Among them those at $t = 0$ and 1 are properly pre-space-moving. Those at $t \geq 2$ are space-moving ones of period 3, which move rightward by one cell in three steps, and hence their speed is $c/3$. Figure 2.7 shows diameter-growing configurations. At $t = 1$ and 2 their diameters decrease, but from $t = 3$ they do not decrease. Actually, the space-moving pattern shown in Fig. 2.6 is generated every three steps, and hence both diameter and population grow indefinitely.

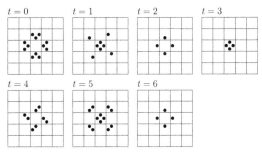

Fig. 2.5   Pre-periodic configurations in irreversible ESPCA-09458f. Configurations at $t = 0$ and 1 are properly pre-periodic, and those at $t \geq 2$ are periodic.

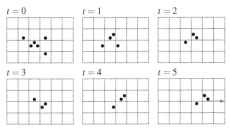

Fig. 2.6   Pre-space-moving configurations in irreversible ESPCA-09458f. Configurations at $t = 0$ and 1 are properly pre-space-moving, and those at $t \geq 2$ are space-moving.

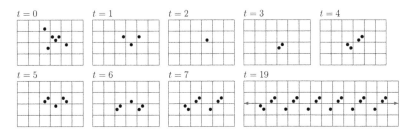

Fig. 2.7   Diameter-growing configurations in irreversible ESPCA-09458f.

Theorem 2.5. *Let P be a reversible ESPCA. If α is a finite configuration of P, then it is either periodic, space-moving, or diameter-growing.*

Proof. Let F be the global function of P. By the same argument in the proof of Theorem 2.4, any finite configuration α is either pre-periodic, pre-space-moving, or diameter-growing.

First, consider the case where α is pre-periodic. Then, there exist integers $m > n \geq 0$ that satisfy $F^m(\alpha) = F^n(\alpha)$. We show that α is periodic in the reversible ESPCA P. Assume, on the contrary, α is not periodic. Then, $\forall t > 0 \ (F^t(\alpha) \neq \alpha)$ holds. From this, $\forall t > 0 \ \forall k \geq 0 \ (F^{t+k}(\alpha) \neq F^k(\alpha))$ follows since F is an injection. Thus, we obtain $F^m(\alpha) \neq F^n(\alpha)$ by letting $t = m - n$ and $k = n$, a contradiction. Therefore, α is periodic.

Second, consider the case where α is pre-space-moving. Then, there exist integers $m > n \geq 0$ and $\mathbf{x}_0 \in D = \mathbb{Z}^2 - \{(0,0)\}$ that satisfy $\forall \mathbf{x} \in \mathbb{Z}^2 \ (F^m(\alpha)(\mathbf{x} + \mathbf{x}_0) = F^n(\alpha)(\mathbf{x}))$. We show that α is space-moving in P. Assume, on the contrary, α is not space-moving. Then, $\forall t > 0 \ \forall \mathbf{x}_0 \in D \ \exists \mathbf{x} \in \mathbb{Z}^2 \ (F^t(\alpha)(\mathbf{x} + \mathbf{x}_0) \neq \alpha(\mathbf{x}))$ holds. From this, $\forall t > 0 \ \forall k \geq 0 \ \forall \mathbf{x}_0 \in D \ \exists \mathbf{x} \in \mathbb{Z}^2 \ (F^{t+k}(\alpha)(\mathbf{x} + \mathbf{x}_0) \neq F^k(\alpha)(\mathbf{x}))$ follows since F is an injection. Thus, we obtain $\forall \mathbf{x}_0 \in D \ \exists \mathbf{x} \in \mathbb{Z}^2 \ (F^m(\alpha)(\mathbf{x} + \mathbf{x}_0) \neq F^n(\alpha)(\mathbf{x}))$ by letting $t = m - n$ and $k = n$, a contradiction. Therefore, α is space-moving.

By above, we can see that in a reversible ESPCA, there is no configuration that is properly pre-periodic or properly pre-space-moving. $\square$

Example 2.3. We consider ESPCA-0945df, whose local function is shown in Fig. 2.8. It has a quiescent state $(0,0,0,0)$. It is a reversible ESPCA, but not a conservative one.

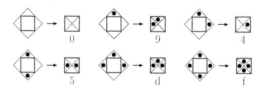

Fig. 2.8   Local function of reversible ESPCA-0945df.

Configurations in Fig. 2.9 are periodic ones of period 6. Configurations in Fig. 2.10 are space-moving ones of period 3, which are the same as the ones in Fig. 2.6. Figure 2.11 shows diameter-growing configurations. At $t = 1$ its diameter decreases, but from $t = 2$ they do not decrease. Actually, two copies of the space-moving pattern shown in Fig. 2.10 are generated every ten steps, and hence both diameter and population grow indefinitely.

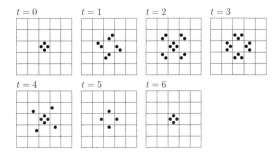

Fig. 2.9   Periodic configurations of period 6 in reversible ESPCA-0945df.

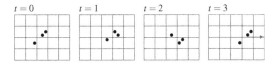

Fig. 2.10   Space-moving configurations of period 3 in reversible ESPCA-0945df.

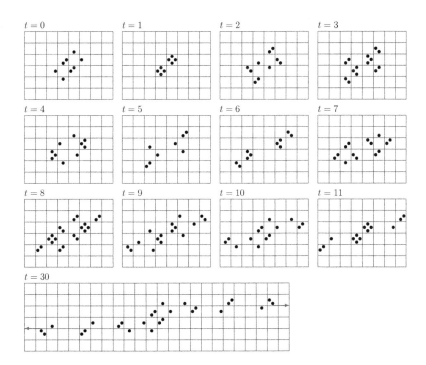

Fig. 2.11   Diameter-growing configurations in reversible ESPCA-0945df.

2.1.5   *Patterns in ESPCA*

As in Sec. 2.1.4, we assume ESPCAs considered in this section have a quiescent state $(0,0,0,0)$. We introduce the notion of a *pattern* as follows. Actually, it is a finite segment of a configuration.

Definition 2.16. Let P be an ESPCA, and D be a finite subset of $\mathbb{Z}^2$. A function $p : D \rightarrow \{0,1\}^4$ is called a *pattern* of P. Let $\alpha \in \mathrm{Conf}_{E4}$, and $\mathbf{x}_0 \in \mathbb{Z}^2$. We say α *contains* the pattern p at $\mathbf{x}_0$ if the following holds.

$$\forall \mathbf{x} \in D \; (\alpha(\mathbf{x} + \mathbf{x}_0) = p(\mathbf{x}))$$

We say α *contains only* the pattern p at $\mathbf{x}_0$ if the following holds.

$$\forall \mathbf{x} \in \mathbb{Z}^2((\mathbf{x} \in D \Rightarrow \alpha(\mathbf{x} + \mathbf{x}_0) = p(\mathbf{x})) \wedge$$
$$(\mathbf{x} \notin D \Rightarrow \alpha(\mathbf{x} + \mathbf{x}_0) = (0,0,0,0)))$$

Usually, when we give a pattern p, we do not specify its domain D, since D can be the set of points whose cells are non-quiescent states.

Definition 2.17. Let P be an ESPCA, and p be a *pattern* of P. Assume a configuration $\alpha \in \mathrm{Conf}_{E4}$ contains only p at $(0,0)$. We say the pattern p is *periodic* (*pre-periodic, space-moving, pre-space-moving,* or *diameter-growing,* respectively), if the configuration α is periodic (pre-periodic, space-moving, pre-space-moving, or diameter-growing). A periodic pattern of period 1 is called a *stable* pattern.

For example, the pattern contained in the configuration at $t = 0$ of Fig. 2.9 is a periodic pattern of period 6. The pattern at $t = 0$ of Fig. 2.10 is a space-moving pattern of period 3.

From Theorems 2.4 and 2.5, we have the following corollaries.

Corollary 2.1. *Let P be an ESPCA, which is generally irreversible. Any pattern p in P is either pre-periodic, pre-space-moving, or diameter-growing.*

Corollary 2.2. *Let P be a reversible ESPCA. Any pattern p in P is either periodic, space-moving, or diameter-growing.*

One of the important tasks in studying reversible ESPCAs is to find useful patterns from which interesting phenomena appear. Small periodic patterns, and space-moving patterns (which are often called *gliders*) are particularly useful, since by interacting these patterns, we can often observe large variety of phenomena that are very interesting or even fantastic.

2.2 Elementary Triangular Partitioned Cellular Automaton

Here, we introduce a triangular partitioned cellular automaton (TPCA), and an elementary TPCA (ETPCA). The reason why we use TPCAs is that its local function is simpler than that of an SPCA, since the number of neighbor cells is only three. In particular, a local function of an ETPCA is described by only four local transition rules, and extremely simple. Nevertheless, it has a high capability of computing as we shall see in Chap. 7. Thus, it is useful for clarifying how computational universality emerges from a simple reversible law.

2.2.1 *Triangular partitioned cellular automaton (TPCA)*

A 3-neighbor triangular partitioned cellular automaton (TPCA) is a PCA whose cell is triangular and is divided into three parts. The cellular space of a TPCA is shown in Fig. 2.12. Though all the cells are identical in their logical operations, there are two kinds of directions, *i.e.*, upward and downward (Fig. 2.13). Therefore, neighbor cells of an up-triangle cell are different from those of a down-triangle cell.

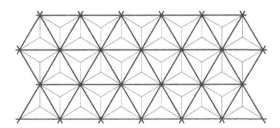

Fig. 2.12   Cellular space of a 3-neighbor TPCA.

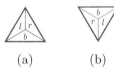

(a)   (b)

Fig. 2.13   (a) Up-triangle cell, and (b) down-triangle cell of a TPCA. In this figure, both of them are in the state (l, b, r).

A local transition rule for an up-triangle cell is depicted in Fig. 2.14(a). The next state of an up-triangle cell is determined depending on the left part of the west neighbor cell, the bottom part of the south cell and the right part of the east cell. On the other hand, for a down-triangle cell, a local transition rule of Fig. 2.14(b) is used. The next state of an down-triangle cell is determined depending on the left part of the east cell, the bottom part of the north cell and the right part of the west cell. In the following, however, we write a local transition rule only of the form of Fig. 2.14(a). We assume up-triangle (down-triangle, respectively) cells are placed at a coordinates $(x, y) \in \mathbb{Z}^2$ such that $x + y$ is even (odd) (see Fig. 2.15).

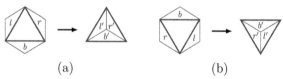

| (a) | (b) |

Fig. 2.14   (a) Local transition rule for up-triangle cells, and (b) that for down-triangle cells of a TPCA. They show the relation $f(l, b, r) = (l', b', r')$ for a local function f.

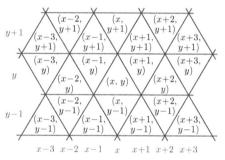

Fig. 2.15   The x-y coordinates in the cellular space of TPCA. If $x + y$ is even, the cell at (x, y) is an up-triangle cell.

Definition 2.18. A *3-neighbor triangular partitioned cellular automaton* (TPCA) is a system defined by

$$P = (\mathbb{Z}^2, (L, B, R), ((-1, 0), (0, -1), (1, 0)), ((1, 0), (0, 1), (-1, 0)), f, \\ (\#_1, \#_2, \#_3)).$$

Here, $\mathbb{Z}^2$ is the set of all 2-dimensional points with integer coordinates at which cells are placed. L, B and R are non-empty finite sets of states of the left, bottom and right parts of a cell. The state set of a cell is thus $Q = L \times B \times R$. The triplet $((-1, 0), (0, -1), (1, 0))$ is a *neighborhood* for a cell at an even parity point (*i.e.*, an up-triangle cell), and $((1, 0), (0, 1), (-1, 0))$ is a neighborhood for a cell at an odd parity point (*i.e.*, a down-triangle cell). The item $f : Q \to Q$ is a *local function*. The state $(\#_1, \#_2, \#_3) \in Q$ is a *quiescent state* (or *blank state*) that satisfies $f(\#_1, \#_2, \#_3) = (\#_1, \#_2, \#_3)$. A TPCA having no quiescent state is also allowed.

If $f(l, b, r) = (l', b', r')$ holds for $(l, b, r), (l', b', r') \in Q$, this relation is called a *local transition rule* of P written pictorially as in Fig. 2.14(a).

Definition 2.19. Let $P = (\mathbb{Z}^2, (L, B, R), ((-1, 0), (0, -1), (1, 0)), ((1, 0), (0, 1), (-1, 0)), f)$ be a TPCA. A *configuration* of P is a function $\alpha : \mathbb{Z}^2 \to Q$. The set of all configurations of P is denoted by $\mathrm{Conf}(P)$, *i.e.*, $\mathrm{Conf}(P) = \{\alpha \mid \alpha : \mathbb{Z}^2 \to Q\}$.

For a TPCA having a quiescent state, we can define the notions of a *support* of a configuration, and *finite* and *infinite configurations* as in the case of SPCA (Definition 2.3).

Definition 2.20. Let $P = (\mathbb{Z}^2, (L, B, R), ((-1, 0), (0, -1), (1, 0)), ((1, 0), (0, 1), (-1, 0)), f, (\#_1, \#_2, \#_3))$ be a TPCA. Let $\alpha \in \mathrm{Conf}(P)$. The set $\mathrm{supp}(\alpha) = \{\mathbf{x} \mid \alpha(\mathbf{x}) \neq (\#_1, \#_2, \#_3)\}$ is called the *support* of α. We say α is a *finite configuration* if $\mathrm{supp}(\alpha)$ is finite. Otherwise, it is called an *infinite configuration*. When $\mathrm{supp}(\alpha) = \emptyset$, α is called a *null configuration*, and is denoted by α_{null}.

For a finite configuration α in a TPCA, we can define its width, height, and diameter, as in the case of an SPCA (Definition 2.4). Thus their definitions are omitted here.

Definition 2.21. Let $P = (\mathbb{Z}^2, (L, B, R), ((-1, 0), (0, -1), (1, 0)), ((1, 0), (0, 1), (-1, 0)), f)$ be a TPCA. Let $\mathrm{pr}_L : Q \to L$ be the *projection function* such that $\mathrm{pr}_L(l, b, r) = l$ for all $(l, b, r) \in Q$. The functions $\mathrm{pr}_B : Q \to B$ and $\mathrm{pr}_R : Q \to R$ are defined similarly. The *global function* $F : \mathrm{Conf}(P) \to \mathrm{Conf}(P)$ of P is defined as the one that satisfies the following.

$\forall \alpha \in \mathrm{Conf}(P), \forall \mathbf{x} \in \mathbb{Z}^2 :$

$$
F(\alpha)(\mathbf{x}) = \begin{cases} f(\mathrm{pr}_L(\alpha(\mathbf{x} + (-1, 0))), \mathrm{pr}_B(\alpha(\mathbf{x} + (0, -1))), \\ \quad \mathrm{pr}_R(\alpha(\mathbf{x} + (1, 0)))) & \text{if } \mathbf{x} \text{ is of even parity} \\ f(\mathrm{pr}_L(\alpha(\mathbf{x} + (1, 0))), \mathrm{pr}_B(\alpha(\mathbf{x} + (0, 1))), \\ \quad \mathrm{pr}_R(\alpha(\mathbf{x} + (-1, 0)))) & \text{if } \mathbf{x} \text{ is of odd parity} \end{cases}
$$

The *evolution process* starting from α is a sequence of configurations $\alpha, F(\alpha), F^2(\alpha), \ldots$.

Note that the argument given in Remark 2.1 is also applied for TPCAs. Hence, the initial configuration should be given using only even parity (or odd parity) cells except stable patterns.

Reversibility of TPCA is defined as in the case of SPCA.

Definition 2.22. Let P be a TPCA whose global function is F. If F is an injection, then P is called a *reversible* TPCA.

The following theorems are proved in a similar manner as in Theorems 2.1 and 2.2 for an SPCA. The difference is that, in a TPCA, its local function is applied using different neighborhood depending on the parity of the point at which a cell is placed. Here, we omit their proof. From these theorems, by designing a TPCA whose local function is injective, we obtain a reversible triangular CA.

Theorem 2.6. *Let P be a TPCA. Its global function F is injective if and only if its local function f is injective.*

Theorem 2.7. *Let P be a TPCA, and F be its global function. Then, there is a triangular CA A having the global function $\hat{F}$ such that $\hat{F} = F$.*

2.2.2 Elementary TPCA (ETPCA)

An elementary triangular partitioned cellular automaton (ETPCA) is a TPCA such that it is rotation-symmetric and each part has only two states. It is an 8-state TPCA. Rotation-symmetry means that if there is a local transition rule of the form shown in Fig. 2.14(a), then there are also rules that are obtained by rotating the both sides of it by a multiple of 60°.

Definition 2.23. Let $P = (\mathbb{Z}^2, (L, B, R), ((-1, 0), (0, -1), (1, 0)), ((1, 0), (0, 1), (-1, 0)), f)$ be a TPCA. The TPCA P is called *rotation-symmetric* if the following conditions (1) and (2) holds.
(1) $L = B = R$
(2) $\forall (l, b, r), (l', b', r') \in L \times D \times R :$
$\quad f(l, b, r) = (l', b', r') \Rightarrow f(b, r, l) = (b', r', l')$

Definition 2.24. Let $P = (\mathbb{Z}^2, (L, B, R), ((-1, 0), (0, -1), (1, 0)), ((1, 0), (0, 1), (-1, 0)), f)$ be a TPCA. The TPCA P is called an *elementary triangular partitioned cellular automaton* (ETPCA), if $L = B = R = \{0, 1\}$, and it is rotation-symmetric. Hence, it is an 8-state TPCA. The set of all *configurations* of an ETPCA is denoted by $\mathrm{Conf}_{E3} = \{\alpha \mid \alpha : \mathbb{Z}^2 \to \{0, 1\}^3\}$.

Since an ETPCA is rotation-symmetric, its local function $f : \{0, 1\}^3 \to \{0, 1\}^3$ is described by only four local transition rules, which are obtained by giving the following four values.

$$f(0, 0, 0), \ f(0, 1, 0), \ f(1, 0, 1), \ f(1, 1, 1)$$

Here, $f(0, 1, 0), f(1, 0, 1) \in \{0, 1\}^3$, but $f(0, 0, 0), f(1, 1, 1) \in \{(0, 0, 0), (1, 1, 1)\}$ since it is rotation-symmetric.

Reading the values of $f(0, 0, 0), f(0, 1, 0), f(1, 0, 1)$ and $f(1, 1, 1)$ as four binary numbers, we can express an ETPCA by a 4-digit octal *identification number $wxyz$* as shown in Fig. 2.16. Thus there are 256 ETPCAs in total.

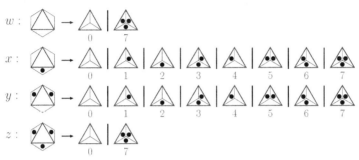

Fig. 2.16　Expressing an ETPCA by a 4-digit octal identification number $wxyz$. Vertical bars indicate alternatives of the right-hand side of each local transition rule.

An ETPCA with the identification number $wxyz$ is denoted by ETPCA-$wxyz$. Its local function and global function are also represented by f_{wxyz} and F_{wxyz}, respectively, using the identification number.

A conservative ETPCA is defined similarly to the case of a conservative ESPCA (Definition 2.9).

Definition 2.25. An ETPCA P is called *conservative* if the number of state 1's (*i.e.*, particles) is conserved in each local transition rule.

From the definitions of reversible ETPCAs and conservative ETPCAs, it is easy to see the following theorem [24].

Theorem 2.8. *Let P be an ETPCA with an ID number $wxyz$.*

(1) *P is reversible if and only if the following condition holds.*
$$(w, z) \in \{(0, 7), (7, 0)\} \wedge (x, y) \in \{1, 2, 4\} \times \{3, 5, 6\} \cup \{3, 5, 6\} \times \{1, 2, 4\}$$
(2) *P is conservative if and only if the following condition holds.*
$$w = 0 \wedge x \in \{1, 2, 4\} \wedge y \in \{3, 5, 6\} \wedge z = 7$$

From the above theorem, we can see that, in the case of ETPCAs, conservative ETPCAs are all reversible. Note that it is not the case in ESPCA (see Theorem 2.3). The total numbers of reversible ETPCAs, conservative ETPCAs, and reversible and conservative ETPCAs are 36, 9, and 9, respectively.

The notion of population for finite configurations of ETPCAs is similarly defined as in the case of ESPCAs (Definition 2.10).

Definition 2.26. The *population* of a cell in the state $(l, b, r) \in \{0, 1\}^3$ is defined by $\mathrm{popul}_3(l, b, r) = l + b + r$. Let $\alpha \in \mathrm{Conf}_{\mathrm{E3}}$ be a finite configuration of an ETPCA. Then, the *population* of α is defined as follows.

$$\mathrm{popul}(\alpha) = \sum_{\mathbf{x} \in \mathrm{supp}(\alpha)} \mathrm{popul}_3(\alpha(\mathbf{x}))$$

Example 2.4. The four local transition rules shown in Fig. 2.17 define the local function f_{0157} of ETPCA-0157. It is easy to see that f_{0157} is injective. Therefore, ETPCA-0157 is reversible. It is also conservative, since each local transition rule conserves the number of the state 1's.

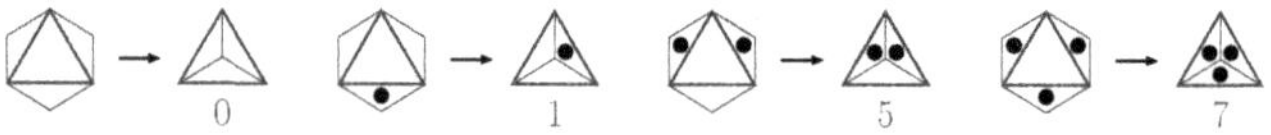

Fig. 2.17   Four local transition rules that define the local function of ETPCA-0157, which is a reversible and conservative ETPCA.

2.2.3   *Evolution of configurations in ETPCA*

In Sec. 2.1.4, the following configurations are defined for ESPCAs having a quiescent state: They are *periodic, pre-periodic, properly pre-periodic, space-moving, pre-space-moving, properly pre-space-moving, diameter-growing, population-growing,* and *population-bounded* configurations. These configurations for ETPCAs and a speed of a space-moving configurations are similarly defined as in Definitions 2.11–2.15.

The following theorems are proved in the same way as in Theorems 2.4 and 2.5. Therefore, we omit their proofs here.

Theorem 2.9. *Let P be an ETPCA, which is generally irreversible. If α is a finite configuration of P, then it is either pre-periodic, pre-space-moving, or diameter-growing.*

Theorem 2.10. *Let P be a reversible ETPCA. If α is a finite configuration of P, then it is either periodic, space-moving, or diameter-growing.*

Example 2.5. Consider ETPCA-0347. Its local function is given in Fig. 2.18. It is a reversible ETPCA having a quiescent state $(0, 0, 0)$.

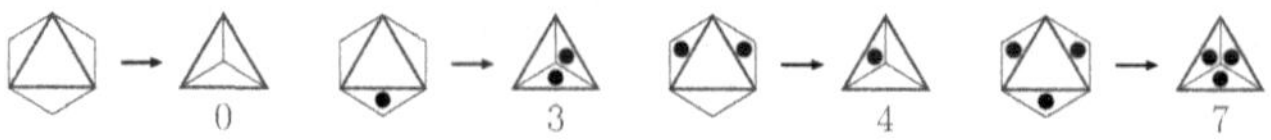

Fig. 2.18   Local function of reversible ETPCA-0347.

Configurations in Fig. 2.19 are periodic ones of period 6. Configurations in Fig. 2.20 are space-moving ones. The space-moving configuration at $t = 0$ shifts rightward by 2 cells in 6 steps. Hence its speed is $c/3$. Figure 2.21 shows diameter-growing configurations. At $t = 2$ two copies of space-moving patterns shown in Fig. 2.20 appear, and they move different directions. Therefore, they are population-bounded configurations as well as diameter-growing ones.

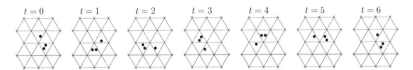

Fig. 2.19   Periodic configurations of period 6 in reversible ETPCA-0347.

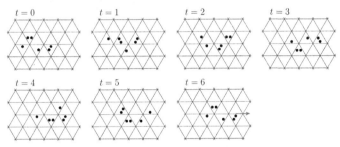

Fig. 2.20   Space-moving configurations of period 6 in reversible ETPCA-0347. The pattern consisting of six particles in each configuration is called a *glider-6*.

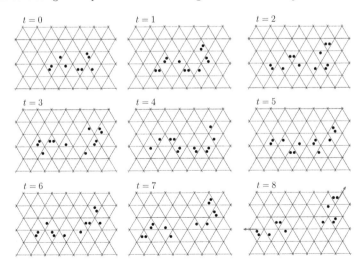

Fig. 2.21   Diameter-growing configurations in reversible ETPCA-0347.

Example 2.6. Consider ETPCA-0340. Its local function is given in Fig. 2.22. It is irreversible, since the first local transition rule has the same right-hand side as the fourth. It has a quiescent state $(0,0,0)$. ETPCA-0340 is similar to ETPCA-0347 (Fig. 2.18), since only their fourth rules are different. In fact, some configurations, *e.g.*, the space-moving ones in Fig. 2.20, evolve exactly in the same way both in these ETPCAs.

Fig. 2.22   Local function of irreversible ETPCA-0340.

Configurations shown in Fig. 2.23 are pre-periodic. Among them, the configurations at $t = 0$ and 1 are properly pre-periodic. Those at $t \geq 2$ are periodic ones of period 6, which are the same periodic configurations shown in Fig. 2.19.

Figure 2.24 shows pre-space-moving configurations. Those at $t = 0$ and 1 are properly pre-space-moving. Those at $t \geq 2$ are space-moving ones of period 6, which are the same ones shown in Fig. 2.20.

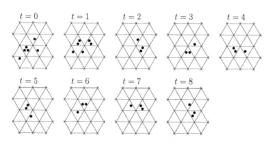

Fig. 2.23   Pre-periodic configurations in irreversible ETPCA-0340. Configurations at $t = 0$ and 1 are properly pre-periodic. Those at $t \geq 2$ are periodic ones.

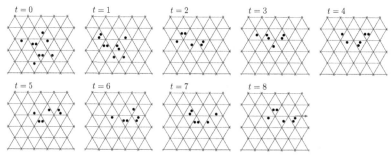

Fig. 2.24   Pre-space-moving configurations in irreversible ETPCA-0340. Configurations at $t = 0$ and 1 are properly pre-space-moving. Those at $t \geq 2$ are space-moving.

Figure 2.25 shows diameter-growing and population-bounded configurations. Similar to the case in Fig. 2.21, two space-moving patterns are generated at $t = 2$, and they move different directions. Note that the evolution process of Fig. 2.21, which differs from that of Fig. 2.25 at $t = 0$ and 1, is also possible in irreversible ETPCA-0340, since the former process does not use the fourth local transition rule.

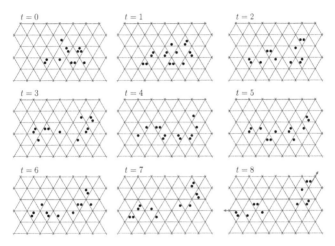

Fig. 2.25   Diameter-growing configurations in irreversible ETPCA-0340.

2.2.4   *Patterns in ETPCA*

In Sec. 2.1.5, the notion of a *pattern* is defined for ESPCAs. Since it is given similarly for ETPCAs, its definition is omitted here.

For example, the pattern at $t = 0$ of Fig. 2.19 is a periodic pattern of period 6, which will be called a *fin*. The pattern at $t = 0$ of Fig. 2.20 is a space-moving pattern of period 6, which will be called a *glider-6*.

From Theorems 2.9 and 2.10, we have the following corollaries.

Corollary 2.3. *Let P be an ETPCA, which is generally irreversible. Any pattern p in P is either pre-periodic, pre-space-moving, or diameter-growing.*

Corollary 2.4. *Let P be a reversible ETPCA. Any pattern p in P is either periodic, space-moving, or diameter-growing.*

As in the case of ESPCAs, it is a key point in the study of reversible ETPCAs to find interesting and useful patterns. We shall see that in some reversible ETPCAs, only from a few kinds of small patterns, reversible computers, like reversible Turing machines, can be systematically constructed despite extreme simplicity of their local functions.

2.3   Remarks and Notes

In this chapter we introduced an SPCA and a TPCA, which are 2-dimensional PCAs having square cells and triangular cells, respectively. The framework of a PCA is useful for designing a reversible CA, because we can obtain a reversible PCA by making its local function injective (Theorems 2.1 and 2.6). ESPCAs and ETPCAs are the simplest subclasses of SPCAs and TPCAs. In the following chapters, we mainly investigate reversible ESPCAs and ETPCAs, and show their computing capabilities, as well as their fascinating behavior.

Two models of reversible ESPCAs were first proposed in [25]. In our current notation they are ESPCA-02c5bf (Fig. 2.3) and ESPCA-02c5df. There, it was proved that a Fredkin gate, a universal reversible logic gate, is realizable in their cellular space. Later, it was shown that reversible Turing machines are realized by implementing a *reversible logic element with memory* (RLEM) (see Sec. 4.3) in ESPCA-01caef [26, 27], and in ESPCAs 01c5ef, 02c5bf and 02c5df [28]. Furthermore, in [29], a model of reversible 81-state SPCA was given, and it was shown that any reversible counter machine, which is a universal computing model, can be implemented as a finite configuration in it (see Chap. 8).

A reversible ETPCA was first proposed in [30]. It is denoted by ETPCA-0157 in our current notation (Fig. 2.17). There, its computational universality was proved by showing that a Fredkin gate is realizable in its cellular space. Later, computational universality of reversible ETPCA-0347 (Fig. 2.18) and reversible ETPCA-0137 was also proved by simulating a Fredkin gate in them [31, 32]. Among them, ETPCA-0347 is the most interesting one, since various fascinating phenomena appear in its cellular space [33] (see Sec. 7.1). Furthermore, by implementing an RLEM, we can compose reversible Turing machines easily in these ETPCAs. Their details are described in Chap. 7.

2.4 Exercises

2.4.1 *Paper-and-pencil exercises*

Exercise 2.1.* Consider ESPCA-0945df (Fig. 2.26).

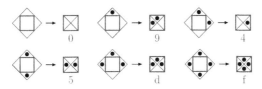

Fig. 2.26   Local function of reversible ESPCA-0945df.

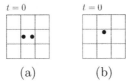

(a)   (b)

Fig. 2.27   Examples of initial configurations of ESPCA-0945df.

(1) Write the evolving process starting from Fig. 2.27(a) for $0 \leq t \leq 6$. Is the configuration periodic, space-moving, or diameter-growing?
(2) Write the evolving process starting from Fig. 2.27(b) for $0 \leq t \leq 7$, and guess how it evolves after that.

Exercise 2.2.* Consider ETPCA-0347 (Fig. 2.28).

Fig. 2.28   Local function of reversible ETPCA-0347.

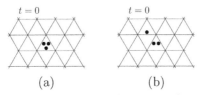

(a)   (b)

Fig. 2.29   Examples of initial configurations of ETPCA-0347.

(1) Write the evolving process starting from Fig. 2.29(a) for $0 \leq t \leq 8$. Is the configuration periodic, space-moving, or diameter-growing?
(2) Write the evolving process starting from Fig. 2.29(b) for $0 \leq t \leq 7$, and infer how it evolves after that.

Exercise 2.3.[**] Suppose P is a reversible ESPCA with a global function F, and α is a diameter-growing configuration of P. Prove that α is also diameter-growing to the negative time direction, *i.e.*, the following holds.

$$\forall d > 0 \; \exists t > 0 \; (\mathrm{diam}((F^{-1})^t(\alpha)) > d)$$

Note that, here, we assume that the total function F^{-1} exists, though it will be proved later in Lemma 3.1.

2.4.2 *Golly exercises*

Exercise 2.4.[*] Consider ESPCA-0945df (Fig. 2.26). Simulate the evolving processes starting from Fig. 2.30(a) and (b) by Golly. Are the configurations periodic, space-moving, or diameter-growing?

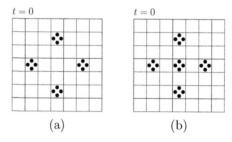

Fig. 2.30   Examples of initial configurations of ESPCA-0945df.

Exercise 2.5.[*] Consider ETPCA-0347 (Fig. 2.28). Simulate the evolving processes starting from Fig. 2.31(a) and (b) by Golly. Are the configurations periodic, space-moving, or diameter-growing?

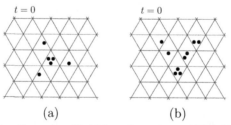

Fig. 2.31   Examples of initial configurations of ETPCA-0347.

Chapter 3

Time-Reversal Symmetry in Reversible PCA

The notion of time-reversal symmetry (T-symmetry) comes from physics. It is the property of a dynamical system where the same evolution law holds for both forward and backward time directions. In this chapter, we investigate how this property is observed in reversible cellular automata. We shall see that the framework of reversible partitioned cellular automata (PCAs) is useful to study it. It is shown that a large number of reversible elementary square PCAs (ESPCAs) and all reversible elementary triangular PCAs (ETPCAs) are T-symmetric under simple transformations on configurations. The obtained results can be used to analyze backward evolution processes, and to design inverse functional modules in reversible PCAs.

3.1 Time-Reversal Symmetry (T-Symmetry)

Time-reversal symmetry (T-symmetry, for short) is a symmetry of an evolution law in a dynamical system under the transformation of reversal of time (see, *e.g.*, a survey paper [34]). Such a property holds for various systems in physics.

Example 3.1. We consider a simple system of the classical mechanics consisting of particles moving in a box as shown in Fig. 3.1. In the configuration at $t = 0$ (*i.e.*, the upper-left one), all the particles are placed at the lower-left corner of the box. The initial velocity of each particle is set randomly. Then, evolve the configuration. If a particle collides with a wall, it bounces. We assume the collision is elastic. At time, say, $t = 100$, it becomes the upper-right configuration, in which particles spread in the box.

If we want to make this evolution process go backward, we first apply the transformation H to the configuration, which reverses velocity vectors of all the particles. By this, we obtain the lower-right configuration. Then, evolve the configuration by the same evolution law as the forward one. At $t' = 0$ all the particles return to the initial positions as shown in the lower-left configuration. Finally, apply again the transformation H that reverses the velocity vectors of all particles. By this, it goes back to the initial configuration, the upper-left one.

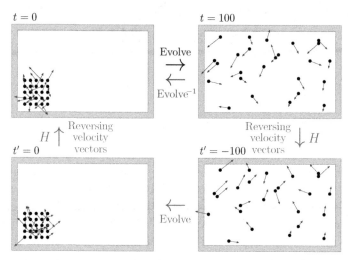

Fig. 3.1   T-symmetry in a simple example system of the classical mechanics. Here, many particles move in a box. Arrows of particles show their velocity vectors.

Do CAs have such a property of T-symmetry? Of course irreversible CAs are not T-symmetric, since their backward evolution is not uniquely determined. In [20, 35–37], T-symmetry of reversible CAs is argued, and it was shown that some kinds of reversible CAs are T-symmetric. In such a CA, its backward evolution is performed by the same global function for the forward one, provided that a simple transformation (similar to H in Fig. 3.1) is applied just before and after applying the global function. For example, the "block CA" of Margolus is T-symmetric in this sense [20, 35].

We study this problem using the framework of PCAs, and show which reversible PCAs are T-symmetric and which transformations are used.

3.2   T-Symmetry in Reversible ESPCA

First, we investigate T-symmetry of reversible ESPCAs. We shall see that about 58% of 1536 reversible ESPCAs, and, in particular, all 128 reversible and conservative ESPCAs are T-symmetric under simple transformations.

3.2.1   *Basic properties of ESPCA*

Before investigating T-symmetries in ESPCAs, we make some preparations. We first define *dualities* among ESPCAs. As we shall see, they have a close relation to T-symmetry. Here, two kinds of dualities are introduced. They are the dualities under *reflection* and *complementation*. These notions are extensions of the ones given in [38] for 1-dimensional elementary cellular automata (ECAs). The dual ESPCAs are essentially the same as the original one in the sense that any evolution process is simulated in the dual ESPCA after applying a simple transformation to the initial configuration.

Definition 3.1. Let P be an ESPCA and $f : \{0,1\}^4 \to \{0,1\}^4$ be its local function. Define $f^{\mathrm{r}} : \{0,1\}^4 \to \{0,1\}^4$ as follows.

$$\forall (t,r,b,l),(t',r',b',l') \in \{0,1\}^4 :$$
$$f(t,r,b,l) = (t',r',b',l') \Leftrightarrow f^{\mathrm{r}}(t,l,b,r) = (t',l',b',r')$$

Then, the ESPCA P^{r} having the local function f^{r} is called the *dual ESPCA of P under reflection*.

From this definition, we can see that the local transition rules of P^{r} are the mirror images of those of P. It means that any evolution process in P is simulated in P^{r} in a straightforward manner by taking the mirror image of the initial configuration (see Lemma 3.2). Note that, in the above definition, the mirror images are taken with respect to the vertical axis (*i.e.*, r and l, and r' and l' are exchanged). However, since ESPCA P is rotation-symmetric (Definition 2.7), it is equivalent to the case where the mirror images are taken with respect to the horizontal axis.

Definition 3.2. Let P be an ESPCA and $f : \{0,1\}^4 \to \{0,1\}^4$ be its local function. For $x \in \{0,1\}$, let $\overline{x} = 1 - x$, *i.e.*, $\overline{x}$ is the complement of x. Define $f^{\mathrm{c}} : \{0,1\}^4 \to \{0,1\}^4$ as follows.

$$\forall (t,r,b,l),(t',r',b',l') \in \{0,1\}^4 :$$
$$f(t,r,b,l) = (t',r',b',l') \Leftrightarrow f^{\mathrm{c}}(\overline{t},\overline{r},\overline{b},\overline{l}) = (\overline{t'},\overline{r'},\overline{b'},\overline{l'})$$

Then, the ESPCA P^{c} having the local function f^{c} is called the *dual ESPCA of P under complementation*.

From this definition, we can see that the local transition rules of P^{c} are obtained from those of P by exchanging 0 and 1. Therefore, any evolution process in P is simulated in P^{c} in a straightforward manner by taking the complement of the initial configuration (see Lemma 3.3).

For an ESPCA P with a local function f, there is an ESPCA P^{rc} whose local function is $(f^{\mathrm{r}})^{\mathrm{c}} = (f^{\mathrm{c}})^{\mathrm{r}}$. It can also be regarded as a kind of a dual ESPCA. We write the local function of P^{rc} by f^{rc} shortly.

Let P be a reversible ESPCA-$uvwxyz$. We denote the identification numbers of the local functions f^{r}_{uvwxyz}, f^{c}_{uvwxyz}, f^{rc}_{uvwxyz}, and f^{-1}_{uvwxyz} by r($uvwxyz$), c($uvwxyz$), rc($uvwxyz$), and inv($uvwxyz$), respectively. Namely, $f^{\mathrm{r}}_{uvwxyz} = f_{\mathrm{r}(uvwxyz)}$, $f^{\mathrm{c}}_{uvwxyz} = f_{\mathrm{c}(uvwxyz)}$, $f^{\mathrm{rc}}_{uvwxyz} = f_{\mathrm{rc}(uvwxyz)}$, and $f^{-1}_{uvwxyz} = f_{\mathrm{inv}(uvwxyz)}$. Note that $f^{-1}_{uvwxyz}(\mathbf{s}') = \mathbf{s}$ if and only if $f_{uvwxyz}(\mathbf{s}) = \mathbf{s}'$ for all $\mathbf{s}, \mathbf{s} \in \{0,1\}^4$.

Example 3.2. Consider the local function $f_{01357\mathrm{f}}$ of ESPCA-01357f. Then, the following holds:

$$f^{\mathrm{r}}_{01357\mathrm{f}} = f_{\mathrm{r}(01357\mathrm{f})} = f_{04357\mathrm{f}}, \qquad f^{\mathrm{c}}_{01357\mathrm{f}} = f_{\mathrm{c}(01357\mathrm{f})} = f_{0235\mathrm{bf}},$$
$$f^{\mathrm{rc}}_{01357\mathrm{f}} = f_{\mathrm{rc}(01357\mathrm{f})} = f_{0235\mathrm{ef}}, \qquad f^{-1}_{01357\mathrm{f}} = f_{\mathrm{inv}(01357\mathrm{f})} = f_{04357\mathrm{f}}.$$

Figure 3.2 shows the above local functions.

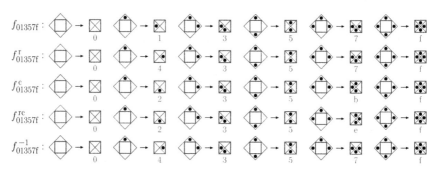

Fig. 3.2   Local function $f_{01357\mathrm{f}}$, and its duals and inverse.

Figure 3.3 gives the list of identification numbers of local functions (f) of 128 reversible and conservative ESPCAs, their dual ones (f^{r}, f^{c} and f^{rc}), and their inverses (f^{-1}). We included inverse local functions besides dual ones in the list, since they will be used in Secs. 3.2.2 and 3.2.3. In each ESPCA, the identification numbers of local functions among $f, f^{\mathrm{r}}, f^{\mathrm{c}}$ and f^{rc} that are equal to f^{-1} are marked by $*$. It will be shown that the ESPCA is T-symmetric under the involutions corresponding to the marked local functions (see Secs. 3.2.2 and 3.2.3).

Note that, since the total number of all reversible ESPCAs is 1536, their complete list is not given in this book. It is found in [39].

f	f^r	f^c	f^{rc}	f^{-1}
01357f	04357f*	0235bf	0235ef	04357f
0135bf	0435ef*	0135bf	0435ef*	0435ef
0135df	0435df*	0835bf	0835ef	0435df
0135ef	0435bf*	0435bf*	0135ef	0435bf
013a7f	043a7f*	023abf	023aef	043a7f
013abf	043aef*	013abf	043aef*	043aef
013adf	043adf*	083abf	083aef	043adf
013aef	043abf*	043abf*	013aef	043abf
01657f	04957f*	0265bf	0295ef	04957f
0165bf	0495ef*	0165bf	0495ef*	0495ef
0165df	0495df*	0865bf	0895ef	0495df
0165ef	0495bf*	0465bf	0195ef	0495bf
016a7f	049a7f*	026abf	029aef	049a7f
016abf	049aef*	016abf	049aef*	049aef
016adf	049adf*	086abf	089aef	049adf
016aef	049abf*	046abf	019aef	049abf
01957f	04657f*	0295bf	0265ef	04657f
0195bf	0465ef*	0195bf	0465ef*	0465ef
0195df	0465df*	0895bf	0865ef	0465df
0195of	0165bf*	0495bf	0165ef	0465bf
019a7f	046a7f*	029abf	026aef	046a7f
019abf	046aef*	019abf	046aef*	046aef
019adf	046adf*	089abf	086aef	046adf
019aef	046abf*	049abf	016aef	046abf
01c57f	04c57f*	02c5bf	02c5ef	04c57f
01c5bf	04c5ef*	01c5bf	04c5ef*	04c5ef
01c5df	04c5df*	08c5bf	08c5ef	04c5df
01c5ef	04c5bf*	04c5bf*	01c5ef	04c5bf
01ca7f	04ca7f*	02cabf	02caef	04ca7f
01cabf	04caef*	01cabf	04caef*	04caef
01cadf	04cadf*	08cabf	08caef	04cadf
01caef	04cabf*	04cabf*	01caef	04cabf
02357f*	02357f*	02357f*	02357f*	02357f
0235bf	0235ef*	01357f	04357f	0235ef
0235df*	0235df*	08357f	08357f	0235df
0235ef	0235bf*	04357f	01357f	0235bf
023a7f*	023a7f*	023a7f*	023a7f*	023a7f
023abf	023aef*	013a7f	043a7f	023aef
023adf*	023adf*	083a7f	083a7f	023adf
023aef	023abf*	043a7f	013a7f	023abf
02657f	02957f*	02657f	02957f*	02957f
0265bf	0295ef*	01657f	04957f	0295ef
0265df	0295df*	08657f	08957f	0295df
0265ef	0295bf*	04657f	01957f	0295bf
026a7f	029a7f*	026a7f	029a7f*	029a7f
026abf	029aef*	016a7f	049a7f	029aef
026adf	029adf*	086a7f	089a7f	029adf
026aef	029abf*	046a7f	019a7f	029abf
02957f	02657f*	02957f	02657f*	02657f
0295bf	0265ef*	01957f	04657f	0265ef
0295df	0265df*	08957f	08657f	0265df
0295ef	0265bf*	04957f	01657f	0265bf
029a7f	026a7f*	029a7f	026a7f*	026a7f
029abf	026aef*	019a7f	046a7f	026aef
029adf	026adf*	089a7f	086a7f	026adf
029aef	026abf*	049a7f	016a7f	026abf
02c57f*	02c57f*	02c57f*	02c57f*	02c57f
02c5bf	02c5ef*	01c57f	04c57f	02c5ef
02c5df*	02c5df*	08c57f	08c57f	02c5df
02c5ef	02c5bf*	04c57f	01c57f	02c5bf
02ca7f*	02ca7f*	02ca7f*	02ca7f*	02ca7f
02cabf	02caef*	01ca7f	04ca7f	02caef
02cadf*	02cadf*	08ca7f	08ca7f	02cadf
02caef	02cabf*	04ca7f	01ca7f	02cabf

f	f^r	f^c	f^{rc}	f^{-1}
04357f	01357f*	0235ef	0235bf	01357f
0435bf	0135ef*	0135ef*	0435bf	0135ef
0435df	0135df*	0835ef	0835bf	0135df
0435ef	0135bf*	0435ef	0135bf*	0135bf
043a7f	013a7f*	023aef	023abf	013a7f
043abf	013aef*	013aef*	043abf	013aef
043adf	013adf*	083aef	083abf	013adf
043aef	013abf*	043aef	013abf*	013abf
04657f	01957f*	0265ef	0295bf	01957f
0465bf	0195ef*	0165ef	0495bf	0195ef
0465df	0195df*	0865ef	0895bf	0195df
0465ef	0195bf*	0465ef	0195bf*	0195bf
046a7f	019a7f*	026aef	029abf	019a7f
046abf	019aef*	016aef	049abf	019aef
046adf	019adf*	086aef	089abf	019adf
046aef	019abf*	046aef	019abf*	019abf
04957f	01657f*	0295ef	0265bf	01657f
0495bf	0165ef*	0195ef	0465bf	0165ef
0495df	0165df*	0895ef	0865bf	0165df
0495ef	0165bf*	0495ef	0165bf*	0165bf
049a7f	016a7f*	029aef	026abf	016a7f
049abf	016aef*	019aef	046abf	016aef
049adf	016adf*	089aef	086abf	016adf
049aef	016abf*	049aef	016abf*	016abf
04c57f	01c57f*	02c5ef	02c5bf	01c57f
04c5bf	01c5ef*	01c5ef*	04c5bf	01c5ef
04c5df	01c5df*	08c5ef	08c5bf	01c5df
04c5ef	01c5bf*	04c5ef	01c5bf*	01c5bf
04ca7f	01ca7f*	02caef	02cabf	01ca7f
04cabf	01caef*	01caef*	04cabf	01caef
04cadf	01cadf*	08caef	08cabf	01cadf
04caef	01cabf*	04caef	01cabf*	01cabf
08357f*	08357f*	0235df	0235df	08357f
0835bf	0835ef*	0135df	0435df	0835ef
0835df*	0835df*	0835df*	0835df*	0835df
0835ef	0835bf*	0435df	0135df	0835bf
083a7f*	083a7f*	023adf	023adf	083a7f
083abf	083aef*	013adf	043adf	083aef
083adf*	083adf*	083adf*	083adf*	083adf
083aef	083abf*	043adf	013adf	083abf
08657f	08957f*	0265df	0295df	08957f
0865bf	0895ef*	0165df	0495df	0895ef
0865df	0895df*	0865df	0895df*	0895df
0865ef	0895bf*	0465df	0195df	0895bf
086a7f	089a7f*	026adf	029adf	089a7f
086abf	089aef*	016adf	049adf	089aef
086adf	089adf*	086adf	089adf*	089adf
086aef	089abf*	046adf	019adf	089abf
08957f	08657f*	0295df	0265df	08657f
0895bf	0865ef*	0195df	0465df	0865ef
0895df	0865df*	0895df	0865df*	0865df
0895ef	0865bf*	0495df	0165df	0865bf
089a7f	086a7f*	029adf	026adf	086a7f
089abf	086aef*	019adf	046adf	086aef
089adf	086adf*	089adf	086adf*	086adf
089aef	086abf*	049adf	016adf	086abf
08c57f*	08c57f*	02c5df	02c5df	08c57f
08c5bf	08c5ef*	01c5df	04c5df	08c5ef
08c5df*	08c5df*	08c5df*	08c5df*	08c5df
08c5ef	08c5bf*	04c5df	01c5df	08c5bf
08ca7f*	08ca7f*	02cadf	02cadf	08ca7f
08cabf	08caef*	01cadf	04cadf	08caef
08cadf*	08cadf*	08cadf*	08cadf*	08cadf
08caef	08cabf*	04cadf	01cadf	08cabf

Fig. 3.3 Identification numbers of 128 reversible and conservative ESPCAs, their duals and inverses [24]. A number with * shows that it is identical to that of f^{-1}.

Next, we define an involution.

Definition 3.3. Let A be a set. A function $H : A \to A$ is called an *involution* if $H \circ H$ is an identity function.

We now define a particular involution $H^{\text{rev}} : \text{Conf}_{\text{E4}} \to \text{Conf}_{\text{E4}}$ by the reversible and conservative ESPCA-08cadf (Fig. 3.4), where Conf_{E4} is the set of all configurations of an ESPCA.

$$H^{\text{rev}} = F_{\text{08cadf}}$$

It is easy to see that $H^{\text{rev}} \circ H^{\text{rev}}$ is an identity function. As it is seen from Fig. 3.4, H^{rev} can be interpreted as the one that reverses the moving directions of all the particles in the cellular space. Thus, H^{rev} is called the *involution of motion reversal*. In the classical mechanics, this operation corresponds to the transformation of the velocity vector $\mathbf{v}$ of each particle to $-\mathbf{v}$ (see Fig. 3.1).

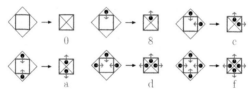

Fig. 3.4 Local function of ESPCA-08cadf by which H^{rev} is defined. H^{rev} is called the involution of motion reversal, since it reverses the moving directions of particles.

The following lemma shows that a backward evolution of reversible ESPCA-$uvwxyz$ is performed by $F_{\text{inv}(uvwxyz)}$ applying H^{rev} just before and after $F_{\text{inv}(uvwxyz)}$. However, it does not mean T-symmetry of the ESPCA, since $F_{\text{inv}(uvwxyz)}$ may be very different from F_{uvwxyz}. In the special case where $F_{\text{inv}(uvwxyz)} = F_{uvwxyz}$ holds, we can say that the backward evolution is carried out by exactly the same global function as the forward one, which we call strict T-symmetry given in Sec. 3.2.2. This lemma is also used for defining weaker T-symmetries as discussed in Sec. 3.2.3.

Lemma 3.1 ([24]). *Let P be a reversible ESPCA-$uvwxyz$ with the local function f_{uvwxyz} and the global function F_{uvwxyz}. Let P' be a reversible ESPCA with the identification number $\text{inv}(uvwxyz)$. The local and global functions of P' are thus $f_{\text{inv}(uvwxyz)} = f^{-1}_{uvwxyz}$ and $F_{\text{inv}(uvwxyz)}$, respectively. Then the following holds.*

$$F^{-1}_{uvwxyz} = H^{\text{rev}} \circ F_{\text{inv}(uvwxyz)} \circ H^{\text{rev}}$$

Proof. Let $\alpha_1 \in \text{Conf}_{E4}$ be any configuration, and $(x_0, y_0) \in \mathbb{Z}^2$ be any point. Let $(t_1, r_1, b_1, l_1) \in \{0, 1\}^4$ be as follows: $\alpha_1(x_0, y_0) = (t_1, r_1, b_1, l_1)$. See Fig. 3.5 that shows the process of state-changes by the operations given below. First, we can see the following relations, where pr_T, pr_R, pr_B, and pr_L are projection functions given in Definition 2.5.

$$\text{pr}_T(H^{\text{rev}}(\alpha_1)(x_0, y_0 - 1)) = b_1$$
$$\text{pr}_R(H^{\text{rev}}(\alpha_1)(x_0 - 1, y_0)) = l_1$$
$$\text{pr}_B(H^{\text{rev}}(\alpha_1)(x_0, y_0 + 1)) = t_1$$
$$\text{pr}_L(H^{\text{rev}}(\alpha_1)(x_0 + 1, y_0)) = r_1$$

Assume $f^{-1}_{uvwxyz}(t_1, r_1, b_1, l_1) = (t_0, r_0, b_0, l_0)$ (*i.e.*, $f_{uvwxyz}(t_0, r_0, b_0, l_0) = (t_1, r_1, b_1, l_1)$). Then, the following holds, since f_{uvwxyz} (and thus f^{-1}_{uvwxyz}) is rotation-symmetric.

$$(F_{\text{inv}(uvwxyz)} \circ H^{\text{rev}}(\alpha_1))(x_0, y_0) = (b_0, l_0, t_0, r_0)$$

Let $\alpha_0 = F_{\text{inv}(uvwxyz)} \circ H^{\text{rev}}(\alpha_1)$. Then, the following relations hold.

$$\text{pr}_T(H^{\text{rev}}(\alpha_0)(x_0, y_0 - 1)) = t_0$$
$$\text{pr}_R(H^{\text{rev}}(\alpha_0)(x_0 - 1, y_0)) = r_0$$
$$\text{pr}_B(H^{\text{rev}}(\alpha_0)(x_0, y_0 + 1)) = b_0$$
$$\text{pr}_L(H^{\text{rev}}(\alpha_0)(x_0 + 1, y_0)) = l_0$$

Hence,

$$(F_{uvwxyz} \circ H^{\text{rev}}(\alpha_0))(x_0, y_0) = (t_1, r_1, b_1, l_1) = \alpha_1(x_0, y_0).$$

By above, the following holds for all $(x_0, y_0) \in \mathbb{Z}^2$.

$$(F_{uvwxyz} \circ H^{\text{rev}} \circ F_{\text{inv}(uvwxyz)} \circ H^{\text{rev}}(\alpha_1))(x_0, y_0) = \alpha_1(x_0, y_0)$$

Thus, $F_{uvwxyz} \circ H^{\text{rev}} \circ F_{\text{inv}(uvwxyz)} \circ H^{\text{rev}}(\alpha_1) = \alpha_1$ for all $\alpha_1 \in \text{Conf}_{E4}$. Therefore,

$$F^{-1}_{uvwxyz} = H^{\text{rev}} \circ F_{\text{inv}(uvwxyz)} \circ H^{\text{rev}}$$

This completes the proof. $\qquad\square$

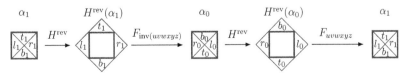

Fig. 3.5   Process of the state-changes around the cell (x_0, y_0) in Lemma 3.1.

3.2.2   Strict T-symmetry in ESPCA

We now define the notion of strict T-symmetry for reversible ESPCAs. It basically follows the definition given in [35].

Definition 3.4. Let P be a reversible ESPCA whose global function is F. If $F^{-1} = H^{\mathrm{rev}} \circ F \circ H^{\mathrm{rev}}$, then P is called *strictly T-symmetric*.

The above definition means that in a strictly T-symmetric reversible ESPCA its backward transition is carried out by exactly the same global function as the one for the forward evolution provided that the moving directions of particles are all reversed before and after the global function is applied. By Lemma 3.1, we have the following theorem.

Theorem 3.1 ([24]). *A reversible ESPCA-uvwxyz is strictly T-symmetric, if* $\mathrm{inv}(uvwxyz) = uvwxyz$. *In this case, the following holds.*

$$F_{uvwxyz}^{-1} = H^{\mathrm{rev}} \circ F_{uvwxyz} \circ H^{\mathrm{rev}}$$

From Fig. 3.3, we can see that there are 16 reversible and conservative ESPCAs such that each of their local function f is identical to its inverse f^{-1}. Therefore, by Theorem 3.1, these 16 ESPCAs are strictly T-symmetric.

Example 3.3. We consider ESPCA-02c5df (Fig. 3.6). It is reversible and conservative. Since $\mathrm{inv}(02\mathrm{c5df}) = 02\mathrm{c5df}$ as shown in Fig. 3.3, it is strictly T-symmetric by Theorem 3.1. Thus the following holds.

$$F_{02\mathrm{c5df}}^{-1} = H^{\mathrm{rev}} \circ F_{02\mathrm{c5df}} \circ H^{\mathrm{rev}}$$

The diagram given in Fig. 3.7 illustrates this relation. It shows that the backward transition from a configuration $\alpha(t+1)$ to $\alpha(t)$ is performed by the same global function $F_{02\mathrm{c5df}}$ for the forward transition. Only the additional operation H^{rev}, which reverses the moving direction of all particles, is required before and after applying $F_{02\mathrm{c5df}}$. We can see that this diagram is analogous to that of Fig. 3.1.

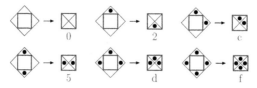

Fig. 3.6   Local function $f_{02\mathrm{c5df}}$ of ESPCA-02c5df.

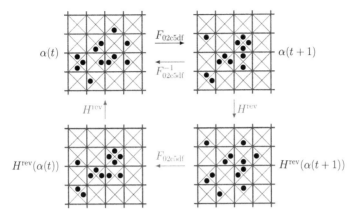

Fig. 3.7   Diagram that illustrates strict T-symmetry of ESPCA-02c5df.

3.2.3   *T-symmetry under a general involution in ESPCA*

Next, we define a weaker version of T-symmetry by replacing the particular involution H^{rev} in Definition 3.4 by an arbitrary involution H. Note that the notion of "weak" T-symmetry in this definition is expressed by the phrase "under the involution H."

Definition 3.5. Let P be a reversible ESPCA whose global function is F. If there is an involution $H : \text{Conf}_{E4} \to \text{Conf}_{E4}$ that satisfies $F^{-1} = H \circ F \circ H$, then P is called *T-symmetric under the involution H*.

We do not restrict the involution H in this definition. However, in this section, we consider the case where H is expressed by $H = H^{\text{rev}} \circ H' = H' \circ H^{\text{rev}}$ for some involution H'. Therefore, $F^{-1} = H^{\text{rev}} \circ H' \circ F \circ H' \circ H^{\text{rev}}$. In this case, the backward evolution is performed by $H' \circ F \circ H'$ applying H^{rev} just before and after $H' \circ F \circ H'$. If H' is a simple involution, we can say that the backward evolution is carried out by a "similar" law as the forward one.

First, we show that ESPCA-$uvwxyz$ which satisfies $\text{inv}(uvwxyz) = \text{r}(uvwxyz)$ is T-symmetric under a certain simple involution. Define a function $\text{refl}_4 : \{0,1\}^4 \to \{0,1\}^4$ as follows: $\text{refl}_4(t,r,b,l) = (t,l,b,r)$ for any $(t,r,b,l) \in \{0,1\}^4$. Next define an involution $H^{\text{refl}} : \text{Conf}_{E4} \to \text{Conf}_{E4}$ as follows. For all $\alpha \in \text{Conf}_{E4}$ and $(x_0, y_0) \in \mathbb{Z}^2$:

$$H^{\text{refl}}(\alpha)(x_0, y_0) = \text{refl}_4(\alpha(-x_0, y_0))$$

It gives the mirror image of a configuration with respect to the y-axis. Thus, H^{refl} is called the *involution of reflection*.

Lemma 3.2 ([24]). *The next relation holds for any ESPCA-uvwxyz.*

$$F_{\text{r}(uvwxyz)} = H^{\text{refl}} \circ F_{uvwxyz} \circ H^{\text{refl}}$$

Proof. First, we show $F_{uvwxyz} = H^{\text{refl}} \circ F_{\text{r}(uvwxyz)} \circ H^{\text{refl}}$. Let $\alpha \in \text{Conf}_{E4}$ be any configuration, and $(x_0, y_0) \in \mathbb{Z}^2$ be any point. Let $(t_0, r_0, b_0, l_0) \in \{0,1\}^4$ be as follows.

$$\begin{aligned}
\text{pr}_T(\alpha(x_0, y_0 - 1)) &= t_0 \\
\text{pr}_R(\alpha(x_0 - 1, y_0)) &= r_0 \\
\text{pr}_B(\alpha(x_0, y_0 + 1)) &= b_0 \\
\text{pr}_L(\alpha(x_0 + 1, y_0)) &= l_0
\end{aligned}$$

See Fig. 3.8 that shows the process of state-changes by the operations given below. In the next step, we apply H^{refl}, and have the following.

$$\begin{aligned}
\text{pr}_T(H^{\text{refl}}(\alpha)(-x_0, y_0 - 1)) &= t_0 \\
\text{pr}_R(H^{\text{refl}}(\alpha)(-x_0 - 1, y_0)) &= l_0 \\
\text{pr}_B(H^{\text{refl}}(\alpha)(-x_0, y_0 + 1)) &= b_0 \\
\text{pr}_L(H^{\text{refl}}(\alpha)(-x_0 + 1, y_0)) &= r_0
\end{aligned}$$

We assume $f_{uvwxyz}(t_0, r_0, b_0, l_0) = (t_1, r_1, b_1, l_1)$. By the definition of duality under reflection, $f_{\text{r}(uvwxyz)}(t_0, l_0, b_0, r_0) = (t_1, l_1, b_1, r_1)$ holds. Thus,

$$(F_{\text{r}(uvwxyz)} \circ H^{\text{refl}}(\alpha))(-x_0, y_0) = (t_1, l_1, b_1, r_1)$$

Finally, we have the following relation for all $\alpha \in \text{Conf}_{E4}$ and (x_0, y_0)

$$\begin{aligned}
(H^{\text{refl}} \circ F_{\text{r}(uvwxyz)} \circ H^{\text{refl}}(\alpha))(x_0, y_0) &= (t_1, r_1, b_1, l_1) \\
&= F_{uvwxyz}(\alpha)(x_0, y_0)
\end{aligned}$$

Therefore, $F_{uvwxyz} = H^{\text{refl}} \circ F_{\text{r}(uvwxyz)} \circ H^{\text{refl}}$ holds, and thus

$$\begin{aligned}
H^{\text{refl}} \circ F_{uvwxyz} \circ H^{\text{refl}} &= H^{\text{refl}} \circ H^{\text{refl}} \circ F_{\text{r}(uvwxyz)} \circ H^{\text{refl}} \circ H^{\text{refl}} \\
&= F_{\text{r}(uvwxyz)}
\end{aligned}$$

This completes the proof. $\square$

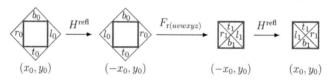

Fig. 3.8   State-changes around the cells (x_0, y_0) and $(-x_0, y_0)$ in Lemma 3.2.

By Lemmas 3.1 and 3.2, we have the following theorem, since it is easy to see $H^{\text{rev}} \circ H^{\text{refl}} = H^{\text{refl}} \circ H^{\text{rev}}$.

Theorem 3.2 ([24]). *A reversible ESPCA-uvwxyz is T-symmetric under the involution $H^{\text{rev}} \circ H^{\text{refl}}$, if inv($uvwxyz$) = r($uvwxyz$). In this case, the following holds.*

$$F_{uvwxyz}^{-1} = H^{\text{rev}} \circ F_{\text{r}(uvwxyz)} \circ H^{\text{rev}}$$
$$= H^{\text{rev}} \circ H^{\text{refl}} \circ F_{uvwxyz} \circ H^{\text{refl}} \circ H^{\text{rev}}$$

From Fig. 3.3, we can see that in *every* reversible and conservative ESPCA its local function f satisfies $f^{-1} = f^{\text{r}}$. Therefore, by Theorem 3.2, *all* the 128 reversible and conservative ESPCAs are T-symmetric under the involution $H^{\text{rev}} \circ H^{\text{refl}}$. Note that there are still many reversible and non-conservative ESPCAs that are T-symmetric under $H^{\text{rev}} \circ H^{\text{refl}}$ as it will be shown in Table 3.1.

Example 3.4. We consider ESPCA-02c5bf (Fig. 3.9). It is reversible and conservative. Since inv(02c5bf) = r(02c5bf) = 02c5ef as shown in Fig. 3.3, it is T-symmetric under the involution $H^{\text{rev}} \circ H^{\text{refl}}$ (Theorem 3.2). Thus the following holds.

$$F_{02c5bf}^{-1} = H^{\text{rev}} \circ F_{\text{r}(02c5bf)} \circ H^{\text{rev}}$$
$$= H^{\text{rev}} \circ H^{\text{refl}} \circ F_{02c5bf} \circ H^{\text{refl}} \circ H^{\text{rev}}$$

The diagram given in Fig. 3.10 illustrates this relation. Namely, the backward transition from a configuration $\alpha(t+1)$ to $\alpha(t)$ is performed by the same global function F_{02c5bf} for the forward transition, provided that the operation $H^{\text{rev}} \circ H^{\text{refl}}$ is applied before and after F_{02c5bf}. In addition, the diagram can also be interpreted that the backward transition is performed by the similar global function $F_{\text{r}(02c5bf)} = F_{02c5ef}$, provided that H^{rev} is applied before and after $F_{\text{r}(02c5bf)}$. Here, "similar" means that each local transition rule for $F_{\text{r}(02c5bf)}$ is a mirror image of the corresponding rule for F_{02c5bf}.

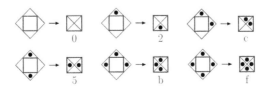

Fig. 3.9   Local function f_{02c5bf} of ESPCA-02c5bf.

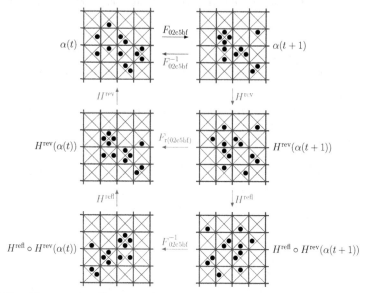

Fig. 3.10   Diagram that illustrates T-symmetry of ESPCA-02c5bf under $H^{\text{rev}} \circ H^{\text{refl}}$.

Next, we show that ESPCA-$uvwxyz$ which satisfies $\text{inv}(uvwxyz) = \text{c}(uvwxyz)$ is T-symmetric under a certain simple involution. Define a function $\text{comp}_4 : \{0,1\}^4 \to \{0,1\}^4$ as follows: $\text{comp}_4(t,r,b,l) = (\bar{t}, \bar{r}, \bar{b}, \bar{l})$ for any $(t,r,b,l) \in \{0,1\}^4$. Then define an involution $H^{\text{comp}} : \text{Conf}_{\text{E4}} \to \text{Conf}_{\text{E4}}$ as follows. For all $\alpha \in \text{Conf}_{\text{E4}}$ and $(x_0, y_0) \in \mathbb{Z}^2$:

$$H^{\text{comp}}(\alpha)(x_0, y_0) = \text{comp}_4(\alpha(x_0, y_0))$$

The involution H^{comp} gives the complement image of a configuration. It is thus called an *involution of complementation*.

Lemma 3.3 ([24]). *The next relation holds for any ESPCA-$uvwxyz$.*

$$F_{\text{c}(uvwxyz)} = H^{\text{comp}} \circ F_{uvwxyz} \circ H^{\text{comp}}$$

Proof. First, we show $F_{uvwxyz} = H^{\text{comp}} \circ F_{\text{c}(uvwxyz)} \circ H^{\text{comp}}$. Let $\alpha \in \text{Conf}_{\text{E4}}$ be any configuration, and $(x_0, y_0) \in \mathbb{Z}^2$ be any point. Let $(t_0, r_0, b_0, l_0) \in \{0,1\}^4$ be as follows.

$$\text{pr}_T(\alpha(x_0, y_0 - 1)) = t_0$$
$$\text{pr}_R(\alpha(x_0 - 1, y_0)) = r_0$$
$$\text{pr}_B(\alpha(x_0, y_0 + 1)) = b_0$$
$$\text{pr}_L(\alpha(x_0 + 1, y_0)) = l_0$$

See Fig. 3.11 that shows the process of state-changes by the operations given below. In the next step, we apply H^{comp}, and have the following.

$$\text{pr}_T(H^{\text{comp}}(\alpha)(x_0, y_0 - 1)) = \overline{t_0}$$
$$\text{pr}_R(H^{\text{comp}}(\alpha)(x_0 - 1, y_0)) = \overline{r_0}$$
$$\text{pr}_B(H^{\text{comp}}(\alpha)(x_0, y_0 + 1)) = \overline{b_0}$$
$$\text{pr}_L(H^{\text{comp}}(\alpha)(x_0 + 1, y_0)) = \overline{l_0}$$

Here, we assume $f_{uvwxyz}(t_0, r_0, b_0, l_0) = (t_1, r_1, b_1, l_1)$. By the definition of duality under complementation, $f_{c(uvwxyz)}(\overline{t_0}, \overline{r_0}, \overline{b_0}, \overline{l_0}) = (\overline{t_1}, \overline{r_1}, \overline{b_1}, \overline{l_1})$ holds. Hence,

$$(F_{c(uvwxyz)} \circ H^{\text{comp}}(\alpha))(x_0, y_0) = (\overline{t_1}, \overline{r_1}, \overline{b_1}, \overline{l_1}).$$

Finally, we have the following relation for all $\alpha \in \text{Conf}_{\text{E4}}$ and (x_0, y_0).

$$(H^{\text{comp}} \circ F_{c(uvwxyz)} \circ H^{\text{comp}}(\alpha))(x_0, y_0) = (t_1, r_1, b_1, l_1)$$
$$= F_{uvwxyz}(\alpha)(x_0, y_0)$$

Therefore, $F_{uvwxyz} = H^{\text{comp}} \circ F_{c(uvwxyz)} \circ H^{\text{comp}}$ holds, and thus

$$H^{\text{comp}} \circ F_{uvwxyz} \circ H^{\text{comp}} = F_{c(uvwxyz)}$$

This completes the proof. □

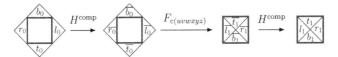

Fig. 3.11   Process of the state-changes around the cell (x_0, y_0) in Lemma 3.3.

By Lemmas 3.1 and 3.3, we have the following theorem, since it is easy to see $H^{\text{rev}} \circ H^{\text{comp}} = H^{\text{comp}} \circ H^{\text{rev}}$.

Theorem 3.3 ([24]). *A reversible ESPCA-uvwxyz is T-symmetric under the involution $H^{\text{rev}} \circ H^{\text{comp}}$, if* inv$(uvwxyz) = $ c$(uvwxyz)$. *In this case, the following holds.*

$$F^{-1}_{uvwxyz} = H^{\text{rev}} \circ F_{c(uvwxyz)} \circ H^{\text{rev}}$$
$$= H^{\text{rev}} \circ H^{\text{comp}} \circ F_{uvwxyz} \circ H^{\text{comp}} \circ H^{\text{rev}}$$

From Theorem 3.3 and Fig. 3.3, we can see that 16 reversible and conservative ESPCAs are T-symmetric under the involution $H^{\text{rev}} \circ H^{\text{comp}}$.

Example 3.5. We consider ESPCA-01c5ef (Fig. 3.12). It is reversible and conservative. Since inv(01c5ef) = c(01c5ef) as shown in Fig. 3.3, it is

T-symmetric under the involution $H^{\text{rev}} \circ H^{\text{comp}}$ (Theorem 3.3). Thus the following holds.

$$F_{01c5ef}^{-1} = H^{\text{rev}} \circ F_{c(01c5ef)} \circ H^{\text{rev}}$$
$$= H^{\text{rev}} \circ H^{\text{comp}} \circ F_{01c5ef} \circ H^{\text{comp}} \circ H^{\text{rev}}$$

The diagram given in Fig. 3.13 illustrates this relation. Namely, the backward transition from a configuration $\alpha(t + 1)$ to $\alpha(t)$ is performed by the global function F_{01c5ef} for the forward transition, provided that the operation $H^{\text{rev}} \circ H^{\text{comp}}$ is applied before and after F_{01c5ef}. The diagram can also be interpreted that the backward transition is performed by the similar global function $F_{c(01c5ef)} = F_{04c5bf}$, provided that H^{rev} is applied before and after $F_{c(01c5ef)}$. Here, "similar" means that each local transition rule for $F_{c(01c5ef)}$ is the 0-1 complementation of the corresponding rule for F_{01c5ef}.

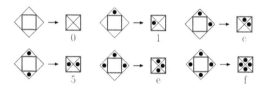

Fig. 3.12   Local function f_{01c5ef} of ESPCA-01c5ef.

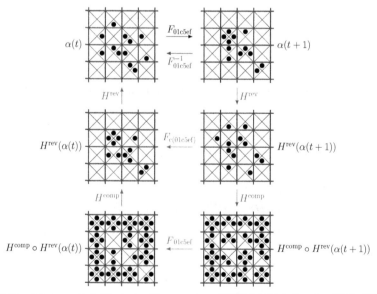

Fig. 3.13   Diagram that illustrates T-symmetry of ESPCA-01c5ef under $H^{\text{rev}} \circ H^{\text{comp}}$.

Combining Lemmas 3.2 and 3.3, we obtain the next lemma.

Lemma 3.4 ([24]). *The next relation holds for any ESPCA-uvwxyz.*

$$F_{rc(uvwxyz)} = H^{refl} \circ H^{comp} \circ F_{uvwxyz} \circ H^{comp} \circ H^{refl}$$

Proof. By Lemma 3.3, $F_{c(uvwxyz)} = H^{comp} \circ F_{uvwxyz} \circ H^{comp}$. Thus

$$\begin{aligned} F_{r(c(uvwxyz))} &= H^{refl} \circ F_{c(uvwxyz)} \circ H^{refl} \\ &= H^{refl} \circ H^{comp} \circ F_{uvwxyz} \circ H^{comp} \circ H^{refl} \end{aligned}$$

holds by Lemma 3.2.   □

By Lemmas 3.1 and 3.4, we have the following theorem.

Theorem 3.4 ([24]). *A reversible ESPCA-uvwxyz is T-symmetric under the involution $H^{rev} \circ H^{refl} \circ H^{comp}$, if $\mathrm{inv}(uvwxyz) = \mathrm{rc}(uvwxyz)$. In this case, the following holds.*

$$\begin{aligned} F^{-1}_{uvwxyz} &= H^{rev} \circ F_{rc(uvwxyz)} \circ H^{rev} \\ &= H^{rev} \circ H^{refl} \circ H^{comp} \circ F_{uvwxyz} \circ H^{comp} \circ H^{refl} \circ H^{rev} \end{aligned}$$

From Theorem 3.4 and Fig. 3.3, we can see 32 reversible and conservative ESPCAs are T-symmetric under the involution $H^{rev} \circ H^{refl} \circ H^{comp}$.

We saw that *every* reversible and conservative ESPCA is T-symmetric under some involution. However, there are many non-conservative ESP-CAs. A complete list of all reversible ESPCAs is given in [39]. Since the number of such ESPCAs is large, we give here only the total numbers of T-symmetric reversible (but may not be conservative) ESPCAs under H^{rev}, $H^{rev} \circ H^{refl}$, $H^{rev} \circ H^{comp}$, and $H^{rev} \circ H^{refl} \circ H^{comp}$ in Table 3.1. The number of non-T-symmetric reversible ESPCAs under these involutions is 640, about 42% of 1536 reversible ESPCAs. It is not known whether each of these ESPCAs becomes T-symmetric under some other involutions.

Table 3.1   Total numbers of reversible ESPCAs, four types of T-symmetric ones, and non-T-symmetric ones [24].

Types of ESPCAs	Numbers
Reversible ESPCAs	1536
T-symmetric reversible ESPCAs under H^{rev} (strictly T-symmetric)	128
T-symmetric reversible ESPCAs under $H^{rev} \circ H^{refl}$	448
T-symmetric reversible ESPCAs under $H^{rev} \circ H^{comp}$	128
T-symmetric reversible ESPCAs under $H^{rev} \circ H^{refl} \circ H^{comp}$	448
Non-T-symmetric reversible ESPCAs under the above involutions	640

Theorems 3.1–3.4 show T-symmetry of one-step evolution of reversible ESPCAs. It is easy to extend the results to many-step evolution.

Lemma 3.5 ([24]). *Let P be a reversible ESPCA-uvwxyz. Assume P is T-symmetric under an involution H, i.e., $F_{uvwxyz}^{-1} = H \circ F_{uvwxyz} \circ H$. Then the following holds for any $n \in \{1, 2, \dots\}$.*

$$(F_{uvwxyz}^{-1})^n = H \circ (F_{uvwxyz})^n \circ H$$

Proof. It is proved by a mathematical induction. The case $n = 1$ is obvious. Assume it holds for $n = k$. Then, $(F_{uvwxyz}^{-1})^{k+1} = H \circ (F_{uvwxyz})^k \circ H \circ H \circ F_{uvwxyz} \circ H = H \circ (F_{uvwxyz})^k \circ F_{uvwxyz} \circ H = H \circ (F_{uvwxyz})^{k+1} \circ H.$ □

We show an application example of Lemma 3.5.

Example 3.6. We consider ESPCA-01caef, whose local function is shown in Fig. 3.14. It is a reversible and conservative ESPCA.

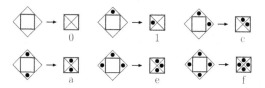

Fig. 3.14   Local function f_{01caef} of ESPCA-01caef.

In ESPCA-01caef, there exist many kinds of space-moving patterns (see Fig. 5.42). Figure 3.15 is one such example having the period 12, which we call here a *glider-12*. It flies in the cellular space in a diagonal direction. As we shall see in Sec. 6.2, any reversible Turing machine is constructed in ESPCA-01caef using the glider-12 as a signal.

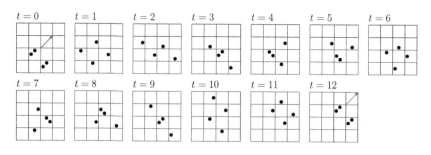

Fig. 3.15   Glider-12, a space-moving pattern of period 12 in ESPCA-01caef.

Figure 3.16 is another example of a space-moving pattern called a *glider-44*, which is of period 44. It moves horizontally or vertically. Here we consider a process that transforms the former glider to the latter. From T-symmetry of ESPCA-01caef, we can obtain its inverse transformation process.

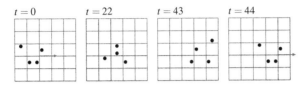

Fig. 3.16   Glider-44, a space-moving pattern of period 44 in ESPCA-01caef.

Consider the configuration $\alpha(0)$ in Fig. 3.17. There are a glider-12 moving to the north-east and a particle. They interact, and produce a glider-44 moving to the east and a particle as shown in $\alpha(145)$. Thus, a glider-12 is converted into a glider-44 by colliding it with a particle.

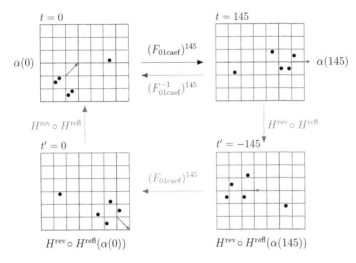

Fig. 3.17   Using the process of converting a glider-12 ($\alpha(0)$) to a glider-44 ($\alpha(145)$), we can convert a glider-44 ($H^{\mathrm{rev}} \circ H^{\mathrm{refl}}(\alpha(145))$) to a glider-12 ($H^{\mathrm{rev}} \circ H^{\mathrm{refl}}(\alpha(0))$) in ESPCA-01caef. It is based on its T-symmetry under $H^{\mathrm{rev}} \circ H^{\mathrm{refl}}$.

ESPCA-01caef is T-symmetric under $H^{\mathrm{rev}} \circ H^{\mathrm{refl}}$ (see Fig. 3.3). Applying $H^{\mathrm{rev}} \circ H^{\mathrm{refl}}$ to $\alpha(145)$, we obtain a configuration that gives the inverse process of the above. In $H^{\mathrm{rev}} \circ H^{\mathrm{refl}}(\alpha(145))$, there are a glider-44 moving to the east, and a particle. Note that the pattern of the glider-44 is the same as the one at $t = 43$ in Fig. 3.16. Namely, the phase of the glider-44 is shifted by the application of $H^{\text{rev}} \circ H^{\text{refl}}$. These two objects interact in ESPCA-01caef, and finally produce a glider-12 moving to the south-east direction and a particle as shown in $H^{\text{rev}} \circ H^{\text{refl}}(\alpha(0))$ of Fig. 3.17. Thus, a glider-44 is converted into a glider-12. Note that the pattern of the glider-12 is the one obtained by rotating the one at $t = 11$ in Fig. 3.15 by 90 degrees clockwise. Namely, the phase and the direction of the glider-12 are changed by the application of $H^{\text{rev}} \circ H^{\text{refl}}$.

Since ESPCA-01caef is T-symmetric under $H^{\text{rev}} \circ H^{\text{comp}}$ as well as $H^{\text{rev}} \circ H^{\text{refl}}$ (see Fig. 3.3), we can obtain a backward evolution process also by this involution, where the glider-12 and glider-44 are represented by "holes," as shown in Fig. 3.18. In this case, however, the resulting configuration is infinite (*i.e.*, it contains an infinite number of non-blank cells). If we want to have a finite configuration that undoes the evolution process of a given finite configuration, this method is not usable.

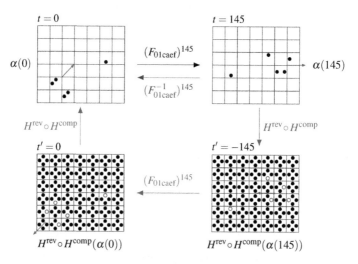

Fig. 3.18   Using the process of converting a glider-12 ($\alpha(0)$) to a glider-44 ($\alpha(145)$), we can convert a complemented glider-44 (shown in $H^{\text{rev}} \circ H^{\text{comp}}(\alpha(145))$ as "holes") to a complemented glider-12 ($H^{\text{rev}} \circ H^{\text{comp}}(\alpha(0))$) in ESPCA-01caef. It is based on its T-symmetry under $H^{\text{rev}} \circ H^{\text{comp}}$. Note that, in $H^{\text{rev}} \circ H^{\text{comp}}(\gamma(0))$ and $H^{\text{rev}} \circ H^{\text{comp}}(\alpha(145))$, the state 0 is represented by a small circle.

3.3 T-Symmetry in Reversible ETPCA

Next, we study T-symmetry of reversible ETPCAs. Definitions on T-symmetries are given in a similar manner as in ESPCAs. We shall see that *all* 36 reversible ETPCAs are T-symmetric under four kinds of simple involutions.

3.3.1 *Basic properties of ETPCA*

As in the case of ESPCA, we consider two kinds of dualities in ETPCA.

Definition 3.6. Let P be an ETPCA and $f : \{0,1\}^3 \to \{0,1\}^3$ be its local function. Define $f^r : \{0,1\}^3 \to \{0,1\}^3$ as follows.

$$\forall (l,b,r), (l',b',r') \in \{0,1\}^3 :$$
$$f(l,b,r) = (l',b',r') \Leftrightarrow f^r(r,b,l) = (r',b',l')$$

Then, the ETPCA P^r having the local function f^r is called the *dual ETPCA of P under reflection.*

Definition 3.7. Let P be an ETPCA and $f : \{0,1\}^3 \to \{0,1\}^3$ be its local function. Let $\bar{x} = 1 - x$ be the complement of x. Define $f^c : \{0,1\}^3 \to \{0,1\}^3$ as follows.

$$\forall (l,b,r), (l',b',r') \in \{0,1\}^3 :$$
$$f(l,b,r) = (l',b',r') \Leftrightarrow f^c(\bar{l},\bar{b},\bar{r}) = (\bar{l'},\bar{b'},\bar{r'})$$

Then, the ETPCA P^c having the local function f^c is called the *dual ETPCA of P under complementation.*

As in the case of ESPCA, for an ETPCA P with a local function f, there is an ETPCA P^{rc} whose local function is $(f^r)^c = (f^c)^r$. We write the local function of P^{rc} by f^{rc} shortly.

We denote the identification numbers of f^r_{wxyz}, f^c_{wxyz}, f^{rc}_{wxyz}, and f^{-1}_{wxyz} by $r(wxyz)$, $c(wxyz)$, $rc(wxyz)$, and $inv(wxyz)$, respectively. Namely, $f^r_{wxyz} = f_{r(wxyz)}$, $f^c_{wxyz} = f_{c(wxyz)}$, $f^{rc}_{wxyz} = f_{rc(wxyz)}$, and $f^{-1}_{wxyz} = f_{inv(wxyz)}$.

Example 3.7. Consider the local function f_{0157} of ETPCA-0157. Then, the following holds:

$$f^r_{0157} = f_{r(0157)} = f_{0457}, \qquad f^c_{0157} = f_{c(0157)} = f_{0267},$$
$$f^{rc}_{0157} = f_{rc(0157)} = f_{0237}, \qquad f^{-1}_{0157} = f_{inv(0157)} = f_{0457}.$$

Figure 3.19 shows the above local functions.

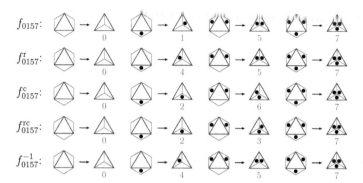

Fig. 3.19   Local function f_{0157}, and its duals and inverse.

f	f^{r}	f^{c}	f^{rc}	f^{-1}		f	f^{r}	f^{c}	f^{rc}	f^{-1}
0137	0467*	0467*	0137	0467		7130	7460*	7460*	7130	7460
0157	0457*	0267	0237	0457		7150	7450*	7260	7230	7450
0167	0437*	0167	0437*	0437		7160	7430*	7160	7430*	7430
0237	0267*	0457	0157	0267		7230	7260*	7450	7150	7260
0257*	0257*	0257*	0257*	0257		7250*	7250*	7250*	7250*	7250
0267	0237*	0157	0457	0237		7260	7230*	7150	7450	7230
0317*	0647	0647	0317*	0317		7310*	7640	7640	7310*	7310
0327	0627	0547	0517*	0517		7320	7620	7540	7510*	7510
0347	0617*	0347	0617*	0617		7340	7610*	7340	7610*	7610
0437	0167*	0437	0167*	0167		7430	7160*	7430	7160*	7160
0457	0157*	0237	0267	0157		7450	7150*	7230	7260	7150
0467	0137*	0137*	0467	0137		7460	7130*	7130*	7460	7130
0517	0547	0627	0327*	0327		7510	7540	7620	7320*	7320
0527*	0527*	0527*	0527*	0527		7520*	7520*	7520*	7520*	7520
0547	0517	0327	0627*	0627		7540	7510	7320	7620*	7620
0617	0347*	0617	0347*	0347		7610	7340*	7610	7340*	7340
0627	0327	0517	0547*	0547		7620	7320	7510	7540*	7540
0647*	0317	0317	0647*	0647		7640*	7310	7310	7640*	7640

Fig. 3.20   Identification numbers of 36 reversible ETPCAs, their dual ones, and inverses [24]. An identification number with * shows that it is identical to that of f^{-1}.

Figure 3.20 shows the list of identification numbers of local functions (f) of all 36 reversible ETPCAs, their dual ones (f^{r}, f^{c} and f^{rc}), and their inverses (f^{-1}).

Let $\mathrm{Conf_{E3}} = \{\alpha \,|\, \alpha : \mathbb{Z}^2 \to \{0,1\}^3\}$ denote the set of all configurations of ETPCA. We define the involution $H_3^{\mathrm{rev}} : \mathrm{Conf_{E3}} \to \mathrm{Conf_{E3}}$ by the reversible ETPCA-0257:

$$H_3^{\mathrm{rev}} = F_{0257}$$

As shown in Figure 3.21, the involution H_3^{rev} is interpreted as the one that reverses the moving directions of all the particles in the cellular space. It is called the *involution of motion reversal*. Though H_3^{rev} is different from H^{rev} for ESPCAs defined in Sec. 3.2.1, it has a similar meaning with the latter. Therefore, in the following, we use the notation H^{rev} in place of H_3^{rev}, since no confusion occurs.

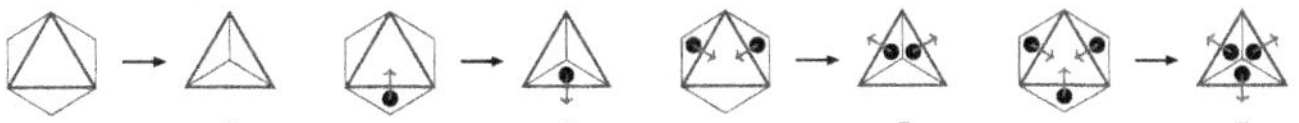

Fig. 3.21   Local function of ETPCA-0257 by which H_3^{rev} is defined. H_3^{rev} is called the involution of motion reversal, since it reverses the moving directions of particles. Hereafter, we use the notation H^{rev} in place of H_3^{rev}.

The following lemma shows the relation between F_{wxyz}^{-1} and $F_{\text{inv}(wxyz)}$, which corresponds to Lemma 3.1 for ESPCAs.

Lemma 3.6 ([24]). *Let P be a reversible ETPCA-wxyz with the local function f_{wxyz} and the global function F_{wxyz}. Let P' be a reversible ETPCA having the identification number $\text{inv}(wxyz)$. Hence, the local and global functions of P' are $f_{\text{inv}(wxyz)} = f_{wxyz}^{-1}$ and $F_{\text{inv}(wxyz)}$, respectively. Then, the following holds.*

$$F_{wxyz}^{-1} = H^{\text{rev}} \circ F_{\text{inv}(wxyz)} \circ H^{\text{rev}}$$

Proof. Let $\alpha_1 \in \text{Conf}_{\text{E3}}$ be any configuration. Let $(x_0, y_0) \in \mathbb{Z}^2$ be any point, and $(l_1, b_1, r_1) \in \{0, 1\}^3$ be as follows: $\alpha_1(x_0, y_0) = (l_1, b_1, r_1)$. See Fig. 3.22 that shows the process of state-changes by the operations given below. We consider only the case where $x_0 + y_0$ is even (*i.e.*, the cell at (x_0, y_0) is an up-triangle), since the other case is similar. First, we can see the following relations.

$$\text{pr}_L(H^{\text{rev}}(\alpha_1)(x_0 - 1, y_0)) = l_1$$
$$\text{pr}_B(H^{\text{rev}}(\alpha_1)(x_0, y_0 - 1)) = b_1$$
$$\text{pr}_R(H^{\text{rev}}(\alpha_1)(x_0 + 1, y_0)) = r_1$$

Assume $f_{wxyz}^{-1}(l_1, b_1, r_1) = (l_0, b_0, r_0)$ (thus, $f_{wxyz}(l_0, b_0, r_0) = (l_1, b_1, r_1)$). Then,

$$(F_{\text{inv}(wxyz)} \circ H^{\text{rev}}(\alpha_1))(x_0, y_0) = (l_0, b_0, r_0).$$

Let $\alpha_0 = F_{\text{inv}(wxyz)} \circ H^{\text{rev}}(\alpha_1)$. Then, the following relations hold.

$$\text{pr}_L(H^{\text{rev}}(\alpha_0)(x_0 - 1, y_0)) = l_0$$
$$\text{pr}_B(H^{\text{rev}}(\alpha_0)(x_0, y_0 - 1)) = b_0$$
$$\text{pr}_R(H^{\text{rev}}(\alpha_0)(x_0 + 1, y_0)) = r_0$$

Therefore,

$$(F_{wxyz} \circ H^{\text{rev}}(\alpha_0))(x_0, y_0) = (l_1, b_1, r_1) = \alpha_1(x_0, y_0)$$

By above, the following holds for all $(x_0, y_0) \in \mathbb{Z}^2$.

$$(F_{wxyz} \circ H^{\text{rev}} \circ F_{\text{inv}(wxyz)} \circ H^{\text{rev}}(\alpha_1))(x_0, y_0) = \alpha_1(x_0, y_0)$$

Thus, $F_{wxyz} \circ H^{\text{rev}} \circ F_{\text{inv}(wxyz)} \circ H^{\text{rev}}(\alpha_1) = \alpha_1$ for all $\alpha_1 \in \text{Conf}_{\text{E3}}$. Hence,

$$F_{wxyz}^{-1} = H^{\text{rev}} \circ F_{\text{inv}(wxyz)} \circ H^{\text{rev}}$$

This completes the proof. □

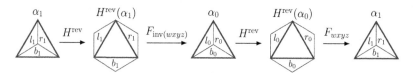

Fig. 3.22   Process of the state-changes around the cell at (x_0, y_0) in Lemma 3.6.

3.3.2   Strict T-symmetry in ETPCA

Definition 3.8. Let P be a reversible ETPCA whose global function is F. If $F^{-1} = H^{\text{rev}} \circ F \circ H^{\text{rev}}$, then P is called *strictly T-symmetric*.

From Lemma 3.6 we have the following theorem.

Theorem 3.5 ([24]). *A reversible ETPCA-wxyz is strictly T-symmetric, if* $\text{inv}(wxyz) = wxyz$. *In this case, the following holds.*

$$F_{wxyz}^{-1} = H^{\text{rev}} \circ F_{wxyz} \circ H^{\text{rev}}$$

From Fig. 3.20 we can see the following.

Corollary 3.1. *The 8 reversible ETPCAs w25z, w31z, w52z and w64z are strictly T-symmetric, where* $(w, z) \in \{(0, 7), (7, 0)\}$.

3.3.3   T-symmetry under a general involution in ETPCA

We now define a weaker version of T-symmetry for reversible ETPCAs.

Definition 3.9. Let P be a reversible ETPCA whose global function is F. If there is an involution $H : \text{Conf}_{\text{E3}} \to \text{Conf}_{\text{E3}}$ that satisfies $F^{-1} = H \circ F \circ H$, then P is called *T-symmetric under the involution* H.

Define a function $\text{refl}_3 : \{0,1\}^3 \to \{0,1\}^3$ as follows: $\text{refl}_3(l,b,r) = (r,b,l)$ for any $(l,b,r) \in \{0,1\}^3$. Next, define an involution H_3^{refl} : $\text{Conf}_{\text{E3}} \to \text{Conf}_{\text{E3}}$ as follows: $H_3^{\text{refl}}(\alpha)(x_0,y_0) = \text{refl}_3(\alpha(-x_0,y_0))$ for all $\alpha \in \text{Conf}_{\text{E3}}$ and $(x_0,y_0) \in \mathbb{Z}^2$. It is called the *involution of reflection*, and gives the *mirror image* of a configuration with respect to the y-axis. As in the case of H_3^{rev}, we hereafter use the notation H^{refl} in place of H_3^{refl}.

Lemma 3.7 ([24]). *The next relation holds for any ETPCA-wxyz.*

$$F_{\text{r}(wxyz)} = H^{\text{refl}} \circ F_{wxyz} \circ H^{\text{refl}}$$

Proof. First, we show $F_{wxyz} = H^{\text{refl}} \circ F_{\text{r}(wxyz)} \circ H^{\text{refl}}$. Let $\alpha \in \text{Conf}_{\text{E3}}$ be any configuration, and $(x_0,y_0) \in \mathbb{Z}^2$ be any point. We consider only the case where $x_0 + y_0$ is even. Let $(l_0,b_0,r_0) \in \{0,1\}^3$ be as follows.

$$\text{pr}_L(\alpha(x_0-1,y_0)) = l_0$$
$$\text{pr}_B(\alpha(x_0,y_0-1)) = b_0$$
$$\text{pr}_R(\alpha(x_0+1,y_0)) = r_0$$

See Fig. 3.23 that shows the process of state-changes by the operations given below. In the next step, we have the following.

$$\text{pr}_L(H^{\text{refl}}(\alpha)(-x_0-1,y_0)) = r_0$$
$$\text{pr}_B(H^{\text{refl}}(\alpha)(-x_0,y_0-1)) = b_0$$
$$\text{pr}_R(H^{\text{refl}}(\alpha)(-x_0+1,y_0)) = l_0$$

Assume $f_{wxyz}(l_0,b_0,r_0) = (l_1,b_1,r_1)$. Since $f_{\text{r}(wxyz)}(r_0,b_0,l_0) = (r_1,b_1,l_1)$,

$$(F_{\text{r}(wxyz)} \circ H^{\text{refl}}(\alpha))(-x_0,y_0) = (r_1,b_1,l_1).$$

Finally, we have the following relation for all α and (x_0,y_0).

$$(H^{\text{refl}} \circ F_{\text{r}(wxyz)} \circ H^{\text{refl}}(\alpha))(x_0,y_0) = (l_1,b_1,r_1) = F_{wxyz}(\alpha)(x_0,y_0)$$

Therefore, $F_{wxyz} = H^{\text{refl}} \circ F_{\text{r}(wxyz)} \circ H^{\text{refl}}$ holds, and thus

$$H^{\text{refl}} \circ F_{wxyz} \circ H^{\text{refl}} = H^{\text{refl}} \circ H^{\text{refl}} \circ F_{\text{r}(wxyz)} \circ H^{\text{refl}} \circ H^{\text{refl}} = F_{\text{r}(wxyz)}$$

This completes the proof. □

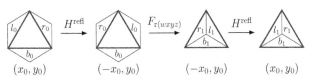

Fig. 3.23   State-changes around the cells at (x_0,y_0) and $(-x_0,y_0)$ in Lemma 3.7.

From Lemmas 3.7 and 3.7 we have the following.

Theorem 3.6 ([24]). *A reversible ETPCA-wxyz is T-symmetric under the involution $H^{\text{refl}} \circ H^{\text{rev}}$, if $\text{inv}(wxyz) = \text{r}(wxyz)$. In this case, the following holds.*

$$
\begin{aligned}
F_{wxyz}^{-1} &= H^{\text{rev}} \circ F_{\text{r}(wxyz)} \circ H^{\text{rev}} \\
&= H^{\text{rev}} \circ H^{\text{refl}} \circ F_{wxyz} \circ H^{\text{refl}} \circ H^{\text{rev}}
\end{aligned}
$$

From Fig. 3.20 we can see the following.

Corollary 3.2. *The 24 reversible ETPCAs $w13z$, $w15z$, $w16z$, $w23z$, $w25z$, $w26z$, $w34z$, $w43z$, $w45z$, $w46z$, $w52z$ and $w61z$ are T-symmetric under the involution $H^{\text{rev}} \circ H^{\text{refl}}$, where $(w, z) \in \{(0, 7), (7, 0)\}$.*

Define a function comp_3 : $\{0, 1\}^3 \rightarrow \{0, 1\}^3$ as follows: $\text{comp}_3(l, b, r) = (\overline{l}, \overline{b}, \overline{r})$ for any $(l, b, r) \in \{0, 1\}^3$. Then, define an involution H_3^{comp}:$\text{Conf}_{\text{E3}} \rightarrow \text{Conf}_{\text{E3}}$ as follows. For all $\alpha \in \text{Conf}_{\text{E3}}$ and $(x_0, y_0) \in \mathbb{Z}^2$:

$$
H_3^{\text{comp}}(\alpha)(x_0, y_0) = \text{comp}_3(\alpha(x_0, y_0))
$$

It is called the *involution of complementation*, and gives the *complement image* of a configuration. Hereafter, we use the notation H^{comp} in place of H_3^{comp}.

Lemma 3.8 ([24]). *The next relation holds for any ETPCA-wxyz.*

$$
F_{\text{c}(wxyz)} = H^{\text{comp}} \circ F_{wxyz} \circ H^{\text{comp}}
$$

Proof. First, we show $F_{wxyz} = H^{\text{comp}} \circ F_{\text{c}(wxyz)} \circ H^{\text{comp}}$. Let $\alpha \in \text{Conf}_{\text{E3}}$ be any configuration, and $(x_0, y_0) \in \mathbb{Z}^2$ be any point. We consider only the case where $x_0 + y_0$ is even. Let $(l_0, b_0, r_0) \in \{0, 1\}^3$ be as follows.

$$
\begin{aligned}
\text{pr}_L(\alpha(x_0 - 1, y_0)) &= l_0 \\
\text{pr}_B(\alpha(x_0, y_0 - 1)) &= b_0 \\
\text{pr}_R(\alpha(x_0 + 1, y_0)) &= r_0
\end{aligned}
$$

See Fig. 3.24 that shows the process of state-changes by the operations given below. In the next step, we have the following.

$$
\begin{aligned}
\text{pr}_L(H^{\text{comp}}(\alpha)(x_0 - 1, y_0)) &= \overline{l_0} \\
\text{pr}_B(H^{\text{comp}}(\alpha)(x_0, y_0 - 1)) &= \overline{b_0} \\
\text{pr}_R(H^{\text{comp}}(\alpha)(x_0 + 1, y_0)) &= \overline{r_0}
\end{aligned}
$$

Assume $f_{wxyz}(l_0, b_0, r_0) = (l_1, b_1, r_1)$. Since $f_{\text{c}(wxyz)}(\overline{l_0}, \overline{b_0}, \overline{r_0}) = (\overline{l_1}, \overline{b_1}, \overline{r_1})$,

$$
(F_{\text{c}(wxyz)} \circ H^{\text{comp}}(\alpha))(x_0, y_0) = (\overline{l_1}, \overline{b_1}, \overline{r_1})
$$

Finally, we have the following relation for all α and (x_0, y_0).

$$(H^{\text{comp}} \circ F_{\text{r}(wxyz)} \circ H^{\text{comp}}(\alpha))(x_0, y_0) = (l_1, b_1, r_1) = F_{wxyz}(\alpha)(x_0, y_0)$$

Therefore, $F_{wxyz} = H^{\text{comp}} \circ F_{\text{c}(wxyz)} \circ H^{\text{comp}}$ holds, and thus

$$H^{\text{comp}} \circ F_{wxyz} \circ H^{\text{comp}} = F_{\text{c}(wxyz)}$$

This completes the proof. $\qquad\qquad\qquad\qquad\qquad\qquad\qquad\qquad$ □

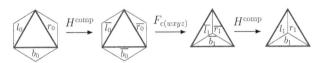

Fig. 3.24   State-changes around the cell (x_0, y_0) in Lemma 3.8.

From Lemmas 3.6 and 3.8 we have the following.

Theorem 3.7 ([24]). *A reversible ETPCA-wxyz is T-symmetric under the involution $H^{\text{rev}} \circ H^{\text{comp}}$, if $\text{inv}(wxyz) = \text{c}(wxyz)$. In this case, the following holds.*

$$\begin{aligned} F_{wxyz}^{-1} &= H^{\text{rev}} \circ F_{\text{c}(wxyz)} \circ H^{\text{rev}} \\ &= H^{\text{rev}} \circ H^{\text{comp}} \circ F_{wxyz} \circ H^{\text{comp}} \circ H^{\text{rev}} \end{aligned}$$

From Fig. 3.20 we can see the following.

Corollary 3.3. *The 8 reversible ETPCAs w13z, w25z, w46z and w52z are T-symmetric under $H^{\text{rev}} \circ H^{\text{comp}}$, where $(w, z) \in \{(0, 7), (7, 0)\}$.*

Combining Lemmas 3.7 and 3.8, we also obtain the next lemma.

Lemma 3.9 ([24]). *The next relation holds for any ETPCA-wxyz.*

$$F_{\text{rc}(wxyz)} = H^{\text{refl}} \circ H^{\text{comp}} \circ F_{wxyz} \circ H^{\text{comp}} \circ H^{\text{refl}}$$

Proof. By Lemma 3.8, we have

$$F_{\text{c}(wxyz)} = H^{\text{comp}} \circ F_{wxyz} \circ H^{\text{comp}}$$

Therefore, by Lemma 3.7, we have

$$F_{\text{r}(\text{c}(wxyz))} = H^{\text{refl}} \circ F_{\text{c}(wxyz)} \circ H^{\text{refl}} = H^{\text{refl}} \circ H^{\text{comp}} \circ F_{wxyz} \circ H^{\text{comp}} \circ H^{\text{refl}}$$

Since $F_{\text{rc}(wxyz)} = F_{\text{r}(\text{c}(wxyz))}$, the lemma holds. $\qquad\qquad\qquad$ □

From Lemmas 3.6 and 3.9 we have the following.

Theorem 3.8 ([24]). *A reversible ETPCA-wxyz is T-symmetric under the involution* $H^{\text{rev}} \circ H^{\text{refl}} \circ H^{\text{comp}}$, *if* $\text{inv}(wxyz) = \text{rc}(wxyz)$. *In this case, the following holds.*

$$F^{-1}_{wxyz} = H^{\text{rev}} \circ F_{\text{rc}(wxyz)} \circ H^{\text{rev}}$$
$$= H^{\text{rev}} \circ H^{\text{refl}} \circ H^{\text{comp}} \circ F_{wxyz} \circ H^{\text{comp}} \circ H^{\text{refl}} \circ H^{\text{rev}}$$

From Fig. 3.20 we can see the following.

Corollary 3.4. *The 24 reversible ETPCAs* $w16z$, $w25z$, $w31z$, $w32z$, $w34z$, $w43z$, $w51z$, $w52z$, $w54z$, $w61z$, $w62z$ *and* $w64z$ *are T-symmetric under the involution* $H^{\text{rev}} \circ H^{\text{refl}} \circ H^{\text{comp}}$, *where* $(w, z) \in \{(0, 7), (7, 0)\}$.

From Corollaries 3.1–3.4, we can see that *every* reversible ETPCA is T-symmetric under either of the involutions H^{rev} (*i.e.*, strictly T-symmetric), $H^{\text{rev}} \circ H^{\text{refl}}$, $H^{\text{rev}} \circ H^{\text{comp}}$, or $H^{\text{rev}} \circ H^{\text{refl}} \circ H^{\text{comp}}$. Table 3.2 shows the numbers of these ETPCAs.

Table 3.2　Total numbers of reversible ETPCAs, four types of T-symmetric ones, and non-T-symmetric ones [24].

Types of ETPCAs	Numbers
Reversible ETPCAs	36
T-symmetric reversible ETPCAs under H^{rev} (strictly T-symmetric)	8
T-symmetric reversible ETPCAs under $H^{\text{rev}} \circ H^{\text{refl}}$	24
T-symmetric reversible ETPCAs under $H^{\text{rev}} \circ H^{\text{comp}}$	8
T-symmetric reversible ESTCAs under $H^{\text{rev}} \circ H^{\text{refl}} \circ H^{\text{comp}}$	24
Non-T-symmetric reversible ETPCAs under the above involutions	0

The following lemma is the ETPCA version of Lemma 3.5. Its proof is omitted here.

Lemma 3.10 ([24]). *Let* P *be a reversible ETPCA with the global function* F_{wxyz}. *Assume* P *is T-symmetric under an involution* H, *i.e.,* $F^{-1}_{wxyz} = H \circ F_{wxyz} \circ H$. *Then the following holds for any* $n \in \{1, 2, \ldots\}$.

$$(F^{-1}_{wxyz})^n = H \circ (F_{wxyz})^n \circ H$$

Below, we give two application examples of Lemma 3.10. They show how T-symmetries in reversible ETPCAs are used to find backward evolution processes.

Example 3.8. Consider ETPCA-0527, whose local function is shown in Fig. 3.25. It is reversible, but not conservative. By Corollary 3.1, it is strictly T-symmetric.

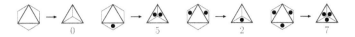

Fig. 3.25   Local function of reversible and non-conservative ETPCA-0527.

If we start from only one particle in ETPCA-0527, an expanding hexagonal pattern is created as shown in Fig. 3.26. At $t = 6$ an isolated particle appears again inside the hexagon. Therefore, a new hexagon is created every 6 steps. By this, concentric hexagons will be formed as shown in $\beta(18)$ in Fig. 3.27.

The process of generating indefinite number of concentric hexagons can be reversed by simply applying H^{rev} to a configuration. As in Fig. 3.27, the configuration $H^{\text{rev}}(\beta(18))$ will become a single particle by applying $(F_{0527})^{18}$. From this, we can see that a one-particle pattern $\beta(0)$ generates concentric hexagons both in the positive and the negative time directions. Note that, since $H^{\text{rev}}(\beta(0))$ is the rotated configuration of $\beta(0)$ by 180 degrees, at $t = -18$ the rotated configuration of $H^{\text{rev}}(\beta(18))$ by 180 degrees appears.

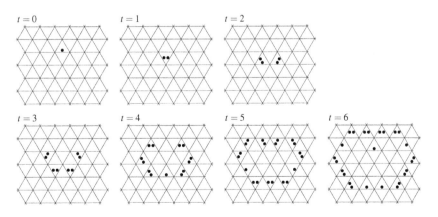

Fig. 3.26   From a one-particle pattern an expanding hexagonal pattern appears in ETPCA-0527. This process is repeated indefinitely, and a large number of concentric hexagons are generated as in $\beta(18)$ of Figure 3.27.

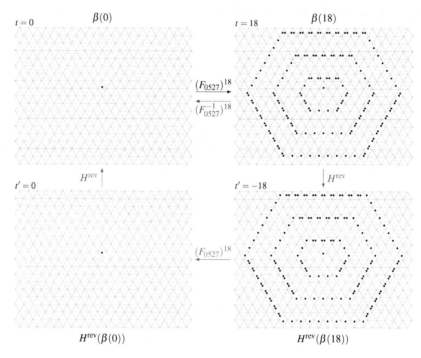

Fig. 3.27   Using the generating process of concentric hexagons ($\beta(18)$) from a one-particle pattern ($\beta(0)$), we can shrink the concentric hexagons ($H^{\mathrm{rev}}(\beta(18))$) to a one-particle pattern ($H^{\mathrm{rev}}(\beta(0))$) in ETPCA-0527. It is based on its strict T-symmetry.

Example 3.9. Consider ETPCA-0347, whose local function is shown in Fig. 3.28. It is a reversible and non-conservative ETPCA. By Corollary 3.2, it is T-symmetric under $H^{\mathrm{rev}} \circ H^{\mathrm{refl}}$.

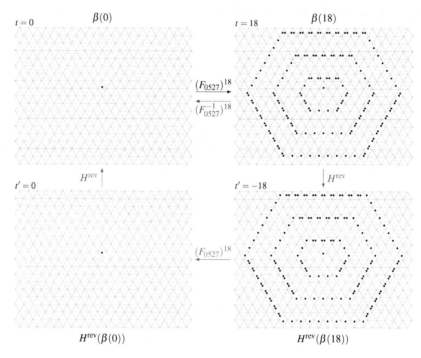

Fig. 3.28   Local function f_{0347} of ETPCA-0347.

As it is already shown in Fig. 2.20, there is a space-moving pattern called a *glider-6* in this cellular space. If we collide two gliders as in $\gamma(0)$ of Fig. 3.29, three gliders are generated after 30 steps ($\gamma(30)$). By this, the number of gliders is increased by one. Its inverse process is obtained by T-symmetry under $H^{\mathrm{rev}} \circ H^{\mathrm{refl}}$. Namely, colliding three gliders as in $H^{\mathrm{rev}} \circ H^{\mathrm{refl}}(\gamma(30))$, we get two gliders ($H^{\mathrm{rev}} \circ H^{\mathrm{refl}}(\gamma(0))$). By this, we obtain a process of decreasing the number of gliders.

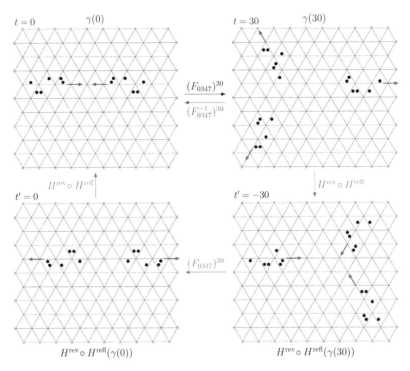

Fig. 3.29   Using the process of generating three gliders ($\gamma(30)$) from two ($\gamma(0)$), we can generate two gliders ($H^{\mathrm{rev}} \circ H^{\mathrm{refl}}(\gamma(0))$) from three ($H^{\mathrm{rev}} \circ H^{\mathrm{refl}}(\gamma(30))$) in ETPCA-0347.

Next, we consider an evolution process starting from a one-particle pattern in ETPCA-0347. Figure 3.30 shows that, if we start from a configuration containing only one particle ($t = 0$), then a disordered pattern appears ($t = 63$), and it expands bigger and bigger ($t = 320$) as if an explosion occurs.

However, if we apply $H^{\mathrm{rev}} \circ H^{\mathrm{refl}}$ to any configuration in the explosion process, it immediately starts to shrink. As seen in Fig. 3.31, the configuration $H^{\mathrm{rev}} \circ H^{\mathrm{refl}}(\delta(64))$ goes to the one-particle configuration $H^{\mathrm{rev}} \circ H^{\mathrm{refl}}(\delta(0))$ after 64 steps. Thus, in ETPCA-0347, both processes of *implosion* and *explosion* to/from a one-particle pattern exist.

We can also observe that a one-particle pattern $\delta(0)$ generates random-like patterns both in the positive and the negative time directions. In fact, if we go to the negative time direction from $\delta(0)$, then at $t = -64$ the configuration obtained by rotating $H^{\mathrm{rev}} \circ H^{\mathrm{refl}}(\delta(64))$ clockwise by 60 degrees will appear.

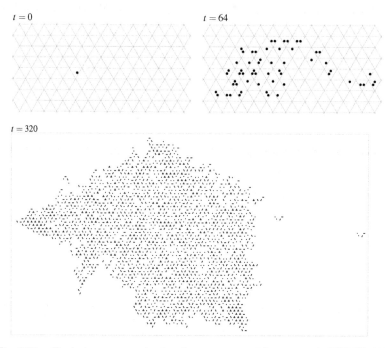

Fig. 3.30   Explosion process starting from a one-particle pattern in ETPCA-0347.

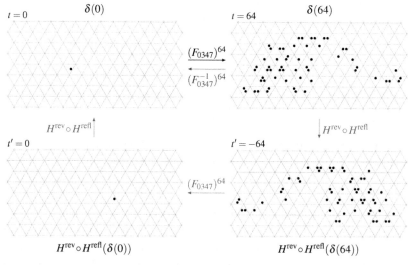

Fig. 3.31   Using the explosion process from a one-particle pattern ($\delta(0)$) to a disordered pattern ($\delta(64)$), we can obtain a implosion process from the disordered pattern ($H^{\text{rev}} \circ H^{\text{refl}}(\delta(64))$) to a one-particle pattern ($H^{\text{rev}} \circ H^{\text{refl}}(\delta(0))$) in ETPCA-0347.

3.4   Remarks and Notes

In this chapter, we investigated T-symmetries in reversible ESPCAs and ETPCAs. The framework of PCAs is useful for formalizing T-symmetries in reversible CAs. This is because the operation corresponding to the transformation of a velocity vector from $\mathbf{v}$ to $-\mathbf{v}$ in the classical mechanics (see Fig. 3.1) is simply expressed by H^{rev} in reversible PCAs. We have shown that 896 reversible ESPCAs among 1536, and all 36 reversible ETPCAs are T-symmetric under simple involutions. However, it is open whether the remaining 640 ESPCAs are T-symmetric under some other involutions.

As shown in Examples 3.6, 3.8 and 3.9, the results on T-symmetry can be used to find and analyze backward evolution processes of reversible PCAs. In the following chapters, T-symmetry will be used to design an "inverse functional module" that undoes the forward function of a given logical or computing module.

T-symmetry in reversible ESPCAs and reversible ETPCAs was studied in [24]. Detailed data on T-symmetry of ESPCAs is found in [39].

3.5   Exercises

3.5.1   *Paper-and-pencil exercise*

Exercise 3.1.[*] Consider ETPCA-0347 (Fig. 3.32). We denote its global function by F_{0347}. ETPCA-0347 is T-symmetric under the involution $\hat{H} = H^{\mathrm{rev}} \circ H^{\mathrm{refl}}$ (Corollary 3.2). Let α be the configuration given in Fig. 3.33. Write the configurations $F_{0347}(\alpha)$, $\hat{H} \circ F_{0347}(\alpha)$, $F_{0347} \circ \hat{H} \circ F_{0347}(\alpha)$, and $\hat{H} \circ F_{0347} \circ \hat{H} \circ F_{0347}(\alpha)$. Observe that $\hat{H} \circ F_{0347} \circ \hat{H} \circ F_{0347}(\alpha) = \alpha$.

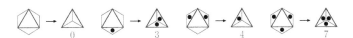

Fig. 3.32   Local function f_{0347} of the reversible ETPCA-0347.

Fig. 3.33   Configuration α in ETPCA-0347.

3.5.2   *Golly exercises*

Exercise 3.2.[**] Consider ESPCA-073a2f (Fig. 3.34). Here, we denote its global function by $\hat{F}$.

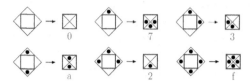

Fig. 3.34   Local function of reversible ESPCA-073a2f.

(1) Show that ESPCA-073a2f is T-symmetric under H^{rev}.
    Note: This is a paper-and-pencil exercise.
(2) Let γ be the single-particle configuration shown in Fig. 3.35. First, get
    the configuration $\hat{F}^{40}(\gamma)$ by simulating ESPCA-073a2f on Golly. Next,
    get $H^{\text{rev}} \circ \hat{F}^{40}(\gamma)$ on Golly. Then, evolve $H^{\text{rev}} \circ \hat{F}^{40}(\gamma)$ in ESPCA-
    073a2f using Golly, and observe its evolution process. In particular,
    observe how the configuration $\hat{F}^{40} \circ H^{\text{rev}} \circ \hat{F}^{40}(\gamma)$ is.

$t = 0$

Fig. 3.35   Single-particle configuration γ in the ESPCA.

Note: To do the above operations on Golly, we use two layers. The
first layer is for simulating ESPCA-073a2f, and the second one is for
ESPCA-08cadf (see Fig. 3.4). If $\hat{F}^{40}(\gamma)$ is obtained on the first layer,
copy the pattern and paste it on the second layer. Since H^{rev} is performed by applying the global function of ESPCA-08cadf, evolve $\hat{F}^{40}(\gamma)$
by one step on the second layer. Then, copy the pattern $H^{\text{rev}} \circ \hat{F}^{40}(\gamma)$,
and paste it on the first layer. Finally, evolve it on the first layer.

Exercise 3.3.[**] Consider ESPCA-07ca2f (Fig. 3.36). Solve the same problems (1) and (2) given in Exercise 3.2.

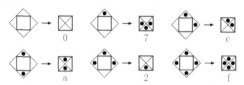

Fig. 3.36   Local function of reversible ESPCA-07ca2f.

Chapter 4

Universal Systems of Reversible Computing

In this chapter, we introduce three models of reversible computing systems besides a reversible cellular automaton (RCA). They are a reversible Turing machine (RTM), a reversible counter machine (RCM), and a reversible logic element with memory (RLEM). An RTM is a standard model of reversible computing, while an RCM is another model simpler than the RTM in some respects. An RLEM is a reversible logic element having different features from reversible logic gates. We first give the notion of Turing universality (or computational universality), and show that restricted classes of RTMs, and the class of RCMs having only two counters are universal. We shall see that RTMs and RCMs can be implemented by a universal RLEM concisely. Behavior of the three models can be viewed using the simulators [21] that run on Golly [1]. They will be helpful to understand how the models work. In the following chapters, we implement RLEMs in several simple reversible PCAs, and then construct RTMs and RCMs out of RLEMs in the PCAs. By this, we can not only obtain universality results of these reversible PCAs, but also observe full computing processes of RTMs and RCMs in their cellular spaces. We also introduce another kind of universality for a CA called intrinsic universality, a property that any CA in some large class of CAs can be simulated by the particular CA. Later (Sec. 6.4.4), it will be shown that if a universal RLEM is implemented in a reversible PCA, then the PCA is intrinsically universal, *i.e.*, it can simulate any reversible PCA.

4.1 Reversible Turing Machine (RTM)

A Turing machine (TM) is a standard model in the theory of computing. According to the Church-Turing thesis [40, 41], any effectively computable function is computed by a TM. In this sense, the class of TMs is considered to be *computationally universal*. Since a TM has a simple structure, it is also convenient to use a reversible version of TM as a standard model in the theory of reversible computing [9]. A reversible TM (RTM) is a "backward deterministic" TM, and hence each computational configuration has at most one predecessor. Lecerf [42] first investigated RTMs, and showed unsolvability of the halting problems and some related problems. Bennett [43] studied them from the viewpoint of thermodynamics of computing, and showed that any irreversible TM can be converted into an equivalent RTM.

4.1.1   *Definitions and examples*

A 1-tape Turing machine (TM) consists of a finite control, a read-write head, and a two-way infinite tape divided into squares in which symbols are written as shown in Fig. 4.1.

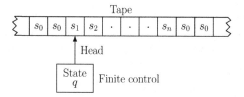

Fig. 4.1   One-tape Turing machine (TM).

Definition 4.1. A *1-tape Turing machine* (TM) is defined by

$$T = (Q, S, q_0, F, s_0, \delta),$$

where Q is a non-empty finite set of states, S is a non-empty finite set of tape symbols, q_0 is an *initial state* ($q_0 \in Q$), F is a set of *final states* ($F \subseteq Q$), and s_0 is a special *blank symbol* ($s_0 \in S$). Here, δ is a move relation, which is a subset of ($Q \times S \times S \times \{L, N, R\} \times Q$). The symbols "$L$," "$N$," and "$R$" are *shift directions* of the head, which stand for "left-shift," "no-shift," and "right-shift," respectively. Each element of δ is a *quintuple* of the form $[p, s, s', d, q]$, which is called a *rule* of T. It means if T reads the symbol s in the state p, then write s', shift the head to the direction d, and go to the state q. We assume each state $q_f \in F$ is a *halting state*, i.e., there is no quintuple of the form $[q_f, s, s', d, q]$ in δ.

Determinism and reversibility of a TM is defined as below.

Definition 4.2. Let $T = (Q, S, q_0, F, s_0, \delta)$ be a TM. We call T a *deterministic TM*, if the following holds for any pair of distinct quintuples

$[p_1, s_1, t_1, d_1, q_1]$ and $[p_2, s_2, t_2, d_2, q_2]$ in δ.

$$(p_1 = p_2) \;\Rightarrow\; (s_1 \neq s_2)$$

It means that for any pair of distinct rules, if the present states are the same, then the read symbols are different.

In this book, we consider only deterministic TMs. Therefore, the term "deterministic" is omitted.

Definition 4.3. Let $T = (Q, S, q_0, F, s_0, \delta)$ be a TM. We call T a *reversible TM* (RTM), if the following holds for any pair of distinct quintuples $[p_1, s_1, t_1, d_1, q_1]$ and $[p_2, s_2, t_2, d_2, q_2]$ in δ.

$$(q_1 = q_2) \;\Rightarrow\; (d_1 = d_2 \;\wedge\; t_1 \neq t_2)$$

It means that for any pair of distinct rules, if the next states are the same, then the shift directions are the same, and the written symbols are different. The above is called the *reversibility condition* for TMs.

In [43], RTMs are defined in a quadruple form, where read-write rules and head-shift rules are separated. This formulation is useful when composing an "inverse" RTM that undoes the computation performed by a given RTM. However, here, we employ the quintuple formulation, since the number of rules for defining an RTM in the quintuple form is about a half of that for defining an RTM in the quadruple form. See Sec. 5.1.3 of [9] for a conversion method between these two forms.

An instantaneous description (ID) of a TM is an expression to describe its finite computational configuration such that the non-blank part of its tape is finite. It is defined as follows.

Definition 4.4. Let $T = (Q, S, q_0, F, s_0, \delta)$ be a TM. We assume $Q \cap S = \emptyset$. An *instantaneous description* (ID) of T is a string of the form $\alpha q \beta$ where $q \in Q$ and $\alpha, \beta \in S^*$. Let λ denote the *empty string*. The ID $\alpha q \beta$ describes the *finite computational configuration* of T where the content of the tape is $\alpha\beta$ (the remaining infinite part of the tape contains only blank symbols), and T is reading the leftmost symbol of β (if $\beta \neq \lambda$) or s_0 (if $\beta = \lambda$) in the state q. An ID $\alpha q \beta$ is called a *standard form ID* if $\alpha \in (S - \{s_0\})S^* \cup \{\lambda\}$, and $\beta \in S^*(S - \{s_0\}) \cup \{\lambda\}$. Namely, a standard form ID is obtained from a general ID by removing superfluous blank symbols from the left and the right ends. An ID $\alpha q_0 \beta$ is called an *initial ID*. An ID $\alpha q \beta$ is called a *final ID* if $q \in F$.

The *transition relation* among standard form IDs of T is denoted by $\vdash_{T}$. Let $\alpha q \beta$ and $\alpha' q' \beta'$ be two standard form IDs. If $\alpha' q' \beta'$ is obtained from $\alpha q \beta$ by applying a rule in δ of T, then we write $\alpha q \beta \vdash_{T} \alpha' q' \beta'$, and say that T goes to the computational configuration $\alpha' q' \beta'$ from $\alpha q \beta$ in one step. For example, if $[q, s, s', R, q'] \in \delta$, $\alpha \in (S - \{s_0\})S^*$, and $\beta \in S^*(S - \{s_0\})$, then $\alpha q s \beta \vdash_{T} \alpha s' q' \beta$. Though the relation $\vdash_{T}$ is conceptually straightforward one, its formal definition is slightly complex, since only standard form IDs are considered, and thus we have to deal with many cases. Hence, its definition is omitted here (see Sect. 5.1.1.3 of [9]).

The reflexive and transitive closure of $\vdash_{T}$ is denoted by $\vdash_{T}^{*}$, which is the transition relation in 0 or more steps. The relation of n-step transition is denoted by $\vdash_{T}^{n}$. Let γ be a standard form ID. We say γ is a *halting ID*, if there is no ID γ' such that $\gamma \vdash_{T} \gamma'$. Let $\alpha_i, \beta_i \in S^*$, and $p_i \in Q$ ($n \in \mathbb{N}, i = 0, 1, \ldots, n$). We say $\alpha_0 p_0 \beta_0 \vdash_{T} \alpha_1 p_1 \beta_1 \vdash_{T} \cdots \vdash_{T} \alpha_n p_n \beta_n$ (or $\alpha_0 p_0 \beta_0 \vdash_{T}^{*} \alpha_n p_n \beta_n$) is a *complete computing process* of T starting from $\alpha_0 p_0 \beta_0$, if $\alpha_0 p_0 \beta_0$ is an initial ID, and $\alpha_n p_n \beta_n$ is a halting ID.

We give three examples of RTMs. Their computing processes can be observed by a simulator [21] constructed on Golly [1].

Example 4.1. An RTM T_{parity} defined below is a very simple example.

$$T_{\text{parity}} = (Q_{\text{parity}}, \{0, 1\}, q_0, \{q_a\}, 0, \delta_{\text{parity}})$$

Here, $Q_{\text{parity}} = \{q_0, q_1, q_2, q_a, q_r\}$, and δ_{parity} are given below.

$$\delta_{\text{parity}} = \{[q_0, 0, 1, R, q_1], \ [q_1, 0, 1, L, q_a], \ [q_1, 1, 0, R, q_2],$$
$$[q_2, 0, 1, L, q_r], \ [q_2, 1, 0, R, q_1]\}$$

It is easy to see that T_{parity} is reversible. Consider the pair of rules $[q_0, 0, 1, R, q_1]$ and $[q_2, 1, 0, R, q_1]$. The next states in these rules are the same (*i.e.*, q_1). We can see the shift directions in them are the same (*i.e.*, R), and the written symbols are different (*i.e.*, 1 and 0). Thus the pair satisfies the reversibility condition in Definition 4.3. No other pair of distinct rules have the same next state. Therefore T_{parity} is reversible. Complete computing processes starting from the IDs $q_0 011$ and $q_0 0111$ are as follows.

$$q_0 011 \ \vdash_{T_{\text{parity}}} 1 q_1 11 \ \vdash_{T_{\text{parity}}} 10 q_2 1 \ \vdash_{T_{\text{parity}}} 100 q_1 \ \vdash_{T_{\text{parity}}} 10 q_a 01$$
$$q_0 0111 \ \vdash_{T_{\text{parity}}} 1 q_1 111 \ \vdash_{T_{\text{parity}}} 10 q_2 11 \ \vdash_{T_{\text{parity}}} 100 q_1 1 \ \vdash_{T_{\text{parity}}} 1000 q_2 \ \vdash_{T_{\text{parity}}} 100 q_r 01$$

For a given string 01^n, the RTM T_{parity} tests whether n is even or not. If it is even, T_{parity} halts in the final (accepting) state q_a. Otherwise it halts in q_r. All the read symbols are complemented.

Example 4.2. An RTM T_{power} is defined by

$$T_{\text{power}} = (Q_{\text{power}}, \{0,1\}, q_0, \{q_a\}, 0, \delta_{\text{power}}).$$

Here, $Q_{\text{power}} = \{q_0, q_1, \ldots, q_7, q_a, q_r\}$, and δ_{power} are given below.

$$\begin{aligned}
\delta_{\text{power}} = \{&[q_0,0,0,R,q_1],\ [q_1,0,0,R,q_2],\ [q_2,0,0,L,q_6],\ [q_2,1,0,R,q_3],\\
&[q_3,0,1,L,q_4],\ [q_3,1,1,R,q_3],\ [q_4,0,0,L,q_7],\ [q_4,1,0,L,q_5],\\
&[q_5,0,1,R,q_2],\ [q_5,1,1,L,q_5],\ [q_6,0,0,L,q_r],\ [q_6,1,1,R,q_1],\\
&[q_7,0,0,L,q_a],\ [q_7,1,1,L,q_r]\}.
\end{aligned}$$

We can verify that T_{power} satisfies the reversibility condition. Complete computing processes starting from $q_0 001111$ and $q_0 00111111$ are as follows.

$$q_0 001111 \quad \left|\frac{31}{T_{\text{power}}}\right. \quad 110\, q_a 1001$$

$$q_0 00111111 \quad \left|\frac{43}{T_{\text{power}}}\right. \quad 111\, q_r 01011\,.$$

For a given string 001^n, the RTM T_{power} tests whether n is a power of 2. If it is the case, T_{power} halts in the final state q_a. Otherwise it halts in q_r. It uses a straightforward algorithm that repeatedly divides the unary number n by 2, and checks the remainder at each division.

Example 4.3. An RTM T_{square} is defined by

$$T_{\text{square}} = (Q_{\text{square}}, \{0,1\}, q_0, \{q_a\}, 0, \delta_{\text{square}}).$$

Here, $Q_{\text{square}} = \{q_0, q_1, \ldots, q_{13}, q_a, q_r\}$, and δ_{square} are given below.

$$\begin{aligned}
\delta_{\text{square}} = \{&[q_0,0,0,R,q_1],\quad [q_1,0,0,R,q_{11}],[q_1,1,1,R,q_1],\quad [q_2,0,0,L,q_8],\\
&[q_2,1,0,L,q_3],\quad [q_3,0,0,L,q_4],\ [q_3,1,1,L,q_3],\quad [q_4,0,1,R,q_5],\\
&[q_4,1,1,L,q_4],\quad [q_5,0,0,L,q_{13}],[q_5,1,0,R,q_6],\quad [q_6,0,0,R,q_7],\\
&[q_6,1,1,R,q_6],\quad [q_7,0,1,L,q_2],\ [q_7,1,1,R,q_7],\quad [q_8,0,0,L,q_a],\\
&[q_8,1,1,R,q_9],\quad [q_9,0,0,R,q_{10}],[q_{10},0,1,R,q_{11}],[q_{10},1,1,R,q_{10}],\\
&[q_{11},0,1,R,q_{12}],[q_{12},0,0,L,q_2],[q_{13},1,0,R,q_r]\ \}.
\end{aligned}$$

We can see that T_{square} satisfies the reversibility condition. Complete computing processes starting from $q_0 01111$ and $q_0 011111$ are as follows.

$$q_0 01111 \quad \left|\frac{63}{T_{\text{square}}}\right. \quad 111\, q_a 100111$$

$$q_0 011111 \quad \left|\frac{102}{T_{\text{square}}}\right. \quad 111110\, q_r 011101\,.$$

For a given string 01^n, the RTM T_{square} tests whether n is a square number, *i.e.*, $n = k^2$ holds for some $k \in \{1, 2, \ldots\}$. If it is the case, T_{square} halts in the final state q_a. Otherwise it halts in q_r. It repeatedly subtract $1, 3, 5, \ldots$ from n, and checks if the result is 0 after each subtraction.

4.1.2 *Turing universality*

Based on the Church-Turing thesis [40, 41], we define the notion of Turing universality for a class $\mathcal{S}$ of computing systems as follows. Here, $\mathcal{S}$ may be either some restricted class of TMs, or a class of computing systems other than TMs, *e.g.*, a class of reversible counter machines (CMs), a class of reversible PCAs, and others.

Definition 4.5. A class $\mathcal{S}$ of computing systems is called *Turing universal* (or *computationally universal*) if for each 1-tape TM there exists a system in $\mathcal{S}$ that simulates it.

Note that the above definition remains unclear, unless we define the notion of *simulation*. Consider the case where a TM simulates some other TM. We say that a TM T' simulates T, if the following holds: For any complete computing process $\alpha \vdash_{T}^{*} \beta$ of T, we can find a complete computing process $\alpha' \vdash_{T'}^{*} \beta'$ of T', such that β is retrieved from β'. Describing the above condition more precisely, the notion of simulation can be defined. The notion of simulation for other models of computing is also defined likewise. However, here, we do not give them for a TM and a CM, since it is cumbersome to describe them. However, the notion of simulation will become intuitively clear from the examples of simulation methods given in this and the following chapters. Their precise definitions are found in [9]: Sec. 5.1.2 for TMs, and Sec. 9.2.2 for CMs. Note that, for CAs, it is given in Sec. 4.4.2 of this book.

The notion of universality in Definition 4.5 is basically given for a *class* of computing systems. However, as noted at the end of Sec. 4.1.3, there is a universal Turing machine (UTM) that can simulate any TM. Therefore, we call such a specific TM universal. Likewise, as we shall see in the later chapters, there is a reversible PCA in which any TM can be simulated. Such a particular PCA is also called Turing universal.

4.1.3 *Turing universality of RTM*

Here, we first explain the result of Bennett [43]: The class of 3-tape RTMs is Turing universal. Then, we show that some very restricted classes of RTMs, *e.g.*, 2-symbol RTMs with a one-way infinite tape, are Turing universal. These results are useful when showing Turing universality of simple reversible PCAs.

Bennett [43] proved that any 1-tape irreversible TM can be converted into an equivalent 3-tape RTM. Hence, the class of 3-tape RTMs is Turing universal.

Theorem 4.1 ([43]). *For any (irreversible) 1-tape TM T, we can construct a reversible 3-tape RTM T' that simulates the former and leaves no garbage information on its tape.*

An outline of the construction method of T' by Bennett is as follows. The first tape of T' simulates the tape of T, which is called a working tape. The second tape is a history tape on which all the history of the computing process of T is recorded step by step. By this, T' reversibly simulates T even if T makes an irreversible movement. When T halts, T' has the same result as T on its working tape. However, garbage information whose amount is proportional to the computing time of T is left on its history tape. It was shown that the garbage information is reversibly erased by the backward computing by the *inverse* RTM. But, before doing so, the computing result on the working tape should be copied on the third tape called an output tape. By above, any TM is simulated by a garbage-less RTM.

Note that there is also a space-efficient method of simulating a TM T by an RTM T'. It has been shown that T' can simulates T using exactly the same number of symbols and exactly the same number of storage tape squares as T under a certain condition (see Sec. 8.2.2 of [9]). In this case, computing time is exponential to the number of used tape squares.

Assume some class of RTMs is known to be Turing universal. If any RTM in this class is simulated by an RTM in another class of RTMs, then the latter class of RTMs is also Turing universal. In this way, Turing universality of various subclasses of RTMs can be shown. In particular, it is possible to show the following.

(1) For any RTM with k two-way infinite tapes, we can construct an RTM with k one-way infinite (*i.e.*, rightward infinite) tapes that simulates the former ($k = 1, 2, \ldots$).

(2) For any RTM with k rightward infinite tapes, we can construct an RTM with only one rightward infinite tape that simulates the former ($k = 2, 3, \ldots$).

(3) For any k-symbol RTM with one rightward infinite tape, we can construct a 2-symbol RTM with one rightward infinite tape that simulates the former ($k = 3, 4, \ldots$).

In the case of irreversible TMs, it is relatively easy to show the results corresponding to the above. However, in the case of RTMs, the simulating TMs should be carefully constructed so that they satisfy the reversibility condition. In addition, the notion of simulation should also be defined properly. These details are found in Sec. 5.3 of [9].

By above, we obtain the following.

Theorem 4.2. *The class of 2-symbol RTMs with a rightward infinite tape is Turing universal.*

By above, to prove Turing universality of a class of computing systems, it suffices to show that they can simulate 2-symbol RTMs with a rightward infinite tape. Also, in the following, when composing RTMs by RLEMs or realizing them in RCAs, we construct only such RTMs.

As for the problem of reducing the number of states of an RTM, it has been shown that a 1-tape many-state RTM can be simulated by a 1-tape 3-state RTM having many symbols (see Sec. 5.3.5 of [9]). Therefore we have the following.

Theorem 4.3. *The class of 3-state RTMs with a rightward infinite tape is Turing universal.*

In the case of irreversible TMs, it is known that a many-state TM can be simulated by a 2-state TM [44]. Hence the class of 2-state TMs is universal. However, it is unknown whether the class of 2-state RTMs is Turing universal.

A *universal Turing machine* (UTM) is one that can simulate any TM. Let UTM(m,n) denote an m-state n-symbol UTM. It is known that various kinds of UTMs with very small m and n exist. For example, Rogozhin [45] gave UTM(4,6), and Neary and Woods [46] gave UTM(6,4), which simulate 2-tag systems and bi-tag systems, respectively. These UTMs have the smallest value of $m \times n$ among the ones so far found.

It is, of course, possible to have a *universal reversible Turing machine* (URTM) by reversifying a UTM using the method of Bennett (Theorem 4.1) and then converting it into a 1-tape RTM. However, if we do so, m and n become very large. In the case of URTM(m,n) with m states and n symbols, a method of simulating cyclic tag systems [47] was used to have ones with small m and n (Sec. 7.3 of [9]). Among them, URTM(10,8) has the smallest value of $m \times n$.

4.2 Reversible Counter Machine (RCM)

A counter machine (CM) is an automaton having a fixed number of counters in which nonnegative integers are stored. Its finite control can increase or decrease the integer stored in each counter, and can test if the integer is zero or not. Minsky [48] proved that any TM can be simulated by a CM with only two counters. Reversible CMs are studied in [49], and it is shown that the class of CMs with two counters is Turing universal even if the reversibility constraint is added. Since such a machine has a simple structure, it is useful to show universality of other reversible systems.

Note that since RCMs are used only in Chap. 8, this section can be skipped until readers go on to Chap. 8.

4.2.1 *Definitions and examples*

A CM is defined as a kind of multi-tape Turing machine as shown in Fig. 4.2. The reason of doing so is that determinism and reversibility are defined similarly to the case of a TM (Definitions 4.2 and 4.3). The tapes are read-only ones, and one-way infinite. The leftmost square of a tape contains the symbol Z, while all the other squares contain P. Therefore, if the machine reads the symbol Z (P, respectively), then it knows the contents of the counter is zero (positive). The increment and decrement operations on a counter are performed by shifting the corresponding head.

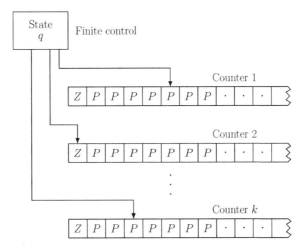

Fig. 4.2 k-counter machine (CM(k)).

Definition 4.6. A *counter machine in the quadruple form* (CM) is defined by

$$M = (Q, k, \delta, q_0, F)$$

Here, Q is a non-empty finite set of states. The integer k ($\in \{1, 2, \ldots\}$) is the number of counters (*i.e.*, tapes). Thus M is also called a *k-counter machine* (CM(k)). The state q_0 is an *initial state* ($q_0 \in Q$), and F is a set of *final states* ($F \subseteq Q$). The CM M uses $\{Z, P\}$ as a tape alphabet. The symbol Z is written only on the leftmost square of each tape, while the symbol P is written in all other squares. The item δ is a move relation, which is a subset of $(Q \times \{1, \ldots, k\} \times \{Z, P\} \times Q) \cup (Q \times \{1, \ldots, k\} \times \{-1, 0, +1\} \times Q)$. The symbols -1, 0 and $+1$ are decrement, no-change and increment operations on a counter, respectively. They are performed by left-shift, no-shift and right-shift operations of a tape head. In the following, -1 and $+1$ are also indicated by $-$ and $+$ for simplicity. Each element of δ is a *quadruple* or a *rule* of the form $[p, i, t, q] \in (Q \times \{1, \ldots, k\} \times \{Z, P\} \times Q)$, or $[p, i, d, q] \in (Q \times \{1, \ldots, k\} \times \{-, 0, +\} \times Q)$. The quadruple $[p, i, t, q]$ is called a *counter-test rule*, and means that if M is in the state p, and the head of the i-th counter reads the symbol t, then go to the state q. The quadruple $[p, i, d, q]$ is called a *count-up/down rule*, and means that if M is in the state p, then shift the i-th head to the direction d by one square, and go to the state q. We assume each state $q_f \in F$ is a *halting state*, *i.e.*, there is no quadruple of the form $[q_f, i, x, q]$ in δ.

Determinism of a CM is defined as follows.

Definition 4.7. Let $M = (Q, k, \delta, q_0, F)$ be a CM. M is called a *deterministic CM*, if the following condition holds for any pair of distinct quadruples $[p_1, i_1, x_1, q_1]$ and $[p_2, i_2, x_2, q_2]$ in δ, where $D = \{-, 0, +\}$.

$$(p_1 = p_2) \Rightarrow (i_1 = i_2 \land x_1 \notin D \land x_2 \notin D \land x_1 \neq x_2)$$

It means that for any pair of distinct rules, if the present states are the same, then they are both counter-test rules on the same counter, and the read symbols are different.

In the following, we deals with only deterministic CMs. Therefore the term "deterministic" is omitted.

Reversibility of a CM is defined as follows.

Definition 4.8. Let $M = (Q, k, \delta, q_0, F)$ be a CM. We call M a *reversible CM* (RCM), if the following holds for any pair of distinct quadruples

$[p_1, i_1, x_1, q_1]$ and $[p_2, i_2, x_2, q_2]$ in δ.

$$(q_1 = q_2) \Rightarrow (i_1 = i_2 \wedge x_1 \notin D \wedge x_2 \notin D \wedge x_1 \neq x_2)$$

It means that for any pair of distinct rules, if the next states are the same, they are both counter-test rules on the same counter, and the read symbols are different. The above is called the *reversibility condition* for CMs.

Next, an instantaneous description (ID) of a CM M, and the transition relation among IDs are defined.

Definition 4.9. Let $M = (Q, k, \delta, q_0, F)$ be a CM. An *instantaneous description* (ID) is an expression $(q, (n_1, n_2, \dots, n_k)) \in Q \times \mathbb{N}^k$. It represents M is in the state q and the i-th counter keeps n_i ($i \in \{1, \dots, k\}$).

Definition 4.10. Let $M = (Q, k, \delta, q_0, F)$ be a CM. The *transition relation* $\vdash_M$ over IDs of M is defined as follows. For every $i \in \{1, \dots, k\}$, $q, q' \in Q$ and $n_1, \dots, n_k, n'_i \in \mathbb{N}$,

$$(q, (n_1, \dots, n_{i-1}, n_i, n_{i+1}, \dots, n_k)) \vdash_M (q', (n_1, \dots, n_{i-1}, n'_i, n_{i+1}, \dots, n_k))$$

holds if and only if one of the following conditions $(1) - (5)$ is satisfied.

(1) $[q, i, Z, q'] \in \delta \wedge n'_i = n_i = 0$.
(2) $[q, i, P, q'] \in \delta \wedge n'_i = n_i > 0$.
(3) $[q, i, -, q'] \in \delta \wedge n'_i = n_i - 1 \geq 0$.
(4) $[q, i, 0, q'] \in \delta \wedge n'_i = n_i$.
(5) $[q, i, +, q'] \in \delta \wedge n'_i = n_i + 1$.

Reflexive and transitive closure of $\vdash_M$ is denoted by $\vdash_M^*$, and n-step transition by $\vdash_M^n$ ($n = 0, 1, \dots$).

An ID $(q, (n_1, n_2, \dots, n_k))$ is called an *initial ID* if $q = q_0$. An ID C is called a *halting ID* if there is no ID C' such that $C \vdash_M C'$. An ID $(q_f, (n_1, \dots, n_k))$ is called a *final ID* if $q_f \in F$. A final ID is a halting ID, since q_f is a halting state. Let C_i ($i \in \{0, 1, \dots, n\}$) be IDs. We say $C_0 \vdash_M C_1 \vdash_M \cdots \vdash_M C_n$ (or $C_0 \vdash_M^* C_n$) is a *complete computing process* of M, if C_0 is an initial ID, and C_n is a final ID.

Example 4.4. We first give a simple example: RCM(2) M_{twice}.

$$M_{\text{twice}} = (\{q_0, q_1, q_2, q_3, q_4, q_5, q_6\}, 2, \delta_{\text{twice}}, q_0, \{q_6\})$$
$$\delta_{\text{twice}} = \{ [q_0, 2, Z, q_1], [q_1, 1, Z, q_6], [q_1, 1, P, q_2], [q_2, 1, -, q_3],$$
$$[q_3, 2, +, q_4], [q_4, 2, +, q_5], [q_5, 2, P, q_1] \}$$

It is easy to verify that M_{twice} is deterministic and reversible. It computes the function $g(x) = 2x$. More precisely, $(q_0, (x, 0)) \vdash^*_{M_{\text{twice}}} (q_6, (0, 2x))$ holds for all $x \in \mathbb{N}$. For example, the complete computing process for $x = 2$ is as follows.

$$(q_0, (2, 0)) \vdash_{M_{\text{twice}}} (q_1, (2, 0)) \vdash_{M_{\text{twice}}} (q_2, (2, 0)) \vdash_{M_{\text{twice}}} (q_3, (1, 0))$$
$$\vdash_{M_{\text{twice}}} (q_4, (1, 1)) \vdash_{M_{\text{twice}}} (q_5, (1, 2)) \vdash_{M_{\text{twice}}} (q_1, (1, 2)) \vdash_{M_{\text{twice}}} (q_2, (1, 2))$$
$$\vdash_{M_{\text{twice}}} (q_3, (0, 2)) \vdash_{M_{\text{twice}}} (q_4, (0, 3)) \vdash_{M_{\text{twice}}} (q_5, (0, 4)) \vdash_{M_{\text{twice}}} (q_1, (0, 4))$$
$$\vdash_{M_{\text{twice}}} (q_6, (0, 4))$$

Example 4.5. Consider the following RCM(3) M_{exp}.

$$M_{\text{exp}} = (\{q_0, q_1, q_2, q_3, q_4, q_5, q_f, t_0, t_1, t_2, t_3, t_4, d_1, d_2, d_3, d_4, d_5\},$$
$$3, \delta_{\text{exp}}, q_0, \{q_f\})$$
$$\delta_{\text{exp}} = \{ [q_0, 2, Z, q_1], \ [q_1, 2, +, q_2], \ [q_2, 1, Z, q_f], \ [q_2, 1, P, t_0],$$
$$[q_3, 1, -, q_4], \ [q_4, 2, -, q_5], \ [q_5, 2, P, q_1],$$
$$[t_0, 3, Z, t_1], \ [t_1, 2, -, t_2], \ [t_2, 3, +, t_3], \ [t_3, 2, Z, d_1],$$
$$[t_3, 2, P, t_4], \ [t_4, 3, P, t_1],$$
$$[d_1, 3, Z, q_3], \ [d_1, 3, P, d_2], \ [d_2, 3, -, d_3], \ [d_3, 2, +, d_4],$$
$$[d_4, 2, +, d_5], \ [d_5, 2, P, d_1] \}$$

We can verify that M_{exp} is deterministic and reversible. M_{exp} computes the function $g(x) = 2^x$. More precisely, the following holds for all $x \in \mathbb{N}$: $(q_0, (x, 0, 0)) \vdash^*_{M_{\text{exp}}} (q_f, (0, 2^x, 0))$. For example, the complete computing process for $x = 3$ is: $(q_0, (3, 0, 0)) \vdash^{84}_{M_{\text{exp}}} (q_f, (0, 8, 0))$. The RCM(3) M_{exp} contains two kinds of "subroutines" M_{trans} and M_{double} in it.

$$M_{\text{trans}} = (\{t_0, t_1, t_2, t_3, t_4, d_1\}, 3, \delta_{\text{trans}}, t_0, \{d_1\})$$
$$\delta_{\text{trans}} = \{ [t_0, 3, Z, t_1], \ [t_1, 2, -, t_2], \ [t_2, 3, +, t_3], \ [t_3, 2, Z, d_1],$$
$$[t_3, 2, P, t_4], \ [t_4, 3, P, t_1] \}$$
$$M_{\text{double}} = (\{d_1, d_2, d_3, d_4, d_5, q_3\}, 3, \delta_{\text{double}}, d_1, \{q_3\})$$
$$\delta_{\text{double}} = \{ [d_1, 3, Z, q_3], \ [d_1, 3, P, d_2], \ [d_2, 3, -, d_3], \ [d_3, 2, +, d_4],$$
$$[d_4, 2, +, d_5], \ [d_5, 2, P, d_1] \}$$

The RCM(3) M_{trans} transfers the contents of the second counter to the third counter, i.e., $(t_0, (x, y, 0)) \vdash^*_{M_{\text{trans}}} (d_1, (x, 0, y))$ holds for all $x \in \mathbb{N}$ and $y \in \mathbb{N} - \{0\}$. The RCM(3) M_{double} doubles the contents of the third counter, and stores it in the second counter, i.e., $(d_1, (x, 0, y)) \vdash^*_{M_{\text{double}}} (q_3, (x, 2y, 0))$ holds for all $x, y \in \mathbb{N}$. Combining M_{trans} and M_{double}, we obtain an RCM(3) that doubles the contents of the second counter. The RCM(3) M_{exp} first sets the second counter to 1. Then it repeatedly decrements the contents of the first counter until it becomes 0, and at each decrement it calls the combined subroutine. By this M_{exp} can compute 2^x.

The above examples are found in the CM simulator contained in [21].

4.2.2 Turing universality of RCM

Minsky [48] showed important results on irreversible CMs. Though his formulation of CM is different from ours, the following results can be proved in essentially the same way.

Lemma 4.1 ([48]). *For any Turing machine T there is a CM(5) M that simulates T.*

Hence, the class of CM(5) is Turing universal. Minsky further showed that any CM(k) ($k \in \{3, 4, \ldots\}$) can be simulated by a CM(2) by a technique of using a *Gödel number*. Namely, the contents of the k counters $n_1, \ldots, n_k$ are encoded into a single integer $p_1^{n_1} \cdots p_k^{n_k}$ called a Gödel number, where p_i is the i-th prime (*i.e.*, $p_1 = 2$, $p_2 = 3$, $p_3 = 5, \ldots$). Thus, the contents of k counters are packed into one counter. One more counter is necessary to perform the operations to these "virtual" counters implemented in the first counter. By this method, CM(2) is shown to be Turing universal.

Lemma 4.2 ([48]). *For any CM(k) M ($k \in \{3, 4, \ldots \}$) there is a CM(2) M' that simulates M.*

From Lemmas 4.1 and 4.2, universality of CM(2) is derived.

Theorem 4.4 ([48]). *The class of CM(2) is Turing universal.*

Now, we investigate Turing universality of *reversible* CMs. Here, we describe only the results shown in [9, 49].

Definition 4.11. Let $M = (Q, k, \delta, q_0, F)$ be a CM. We say M is in the *normal form*, if the initial state q_0 does not appear as the fourth component of any rule in δ (hence q_0 appears only at time $t = 0$), and F is a singleton.

The RCMs M_{twice} and M_{exp} in Examples 4.4 and 4.5 are in the normal form. It is easy to show the next lemma, since M' can be irreversible.

Lemma 4.3 ([9]). *For any CM M, we can construct a CM M' in the normal form that simulates M.*

The following lemma shows that any irreversible CM with k counters is simulated by a reversible CM without leaving garbage information by increasing the number of counters to $2k + 2$. It is proved based on the method of Bennett [43] for RTMs (see Theorem 4.1).

Lemma 4.4 ([9, 49]). *For any CM(k) $M = (Q, k, \delta, q_0, \{q_f\})$ in the normal form, there is an RCM(2k + 2) $M^\dagger = (Q^\dagger, 2k + 2, \delta^\dagger, q_0, \{\hat{q}_0\})$ in the normal form that satisfies the following.*

$$\forall m_1, \ldots, m_k, n_1, \ldots, n_k \in \mathbb{N}$$

$$((q_0, (m_1, \ldots, m_k)) \vdash_{\overline{M}}^* (q_f, (n_1, \ldots, n_k)) \leftrightarrow$$

$$(q_0, (m_1, \ldots, m_k, 0, 0, 0, \ldots, 0)) \vdash_{\overline{M^\dagger}}^* (\hat{q}_0, (m_1, \ldots, m_k, 0, 0, n_1, \ldots, n_k))$$

The next lemma shows that any RCM with k counters can be simulated by an RCM with only two counters.

Lemma 4.5 ([9, 49]). *For any RCM(k) $M = (Q, k, \delta, q_0, \{q_f\})$ in the normal form ($k \in \{3, 4, \ldots\}$), there is an RCM(2) $M^\dagger = (Q^\dagger, 2, \delta^\dagger, q_0, \{q_f\})$ in the normal form that satisfies the following, where p_i is the i-th prime number.*

$$\forall m_1, \ldots, m_k, n_1, \ldots, n_k \in \mathbb{N}$$

$$((q_0, (m_1, \ldots, m_k)) \vdash_{\overline{M}}^* (q_f, (n_1, \ldots, n_k)) \iff$$

$$(q_0, (p_1^{m_1} \cdots p_k^{m_k}, 0)) \vdash_{\overline{M^\dagger}}^* (q_f, (p_1^{n_1} \cdots p_k^{n_k}, 0))$$

From Lemmas 4.1, 4.4 and 4.5, we can conclude that RCMs with only two counters are universal.

Theorem 4.5 ([9, 49]). *The class of RCM(2) is Turing universal.*

4.2.3 RCM in the program form

We newly introduce another formulation of CMs. In Chap. 8, RCMs are constructed in a reversible PCA. There, each rule, which is represented by a quadruple, is further decomposed into a few simpler instructions. A whole RCM is thus given by a sequence of instructions, a kind of a program. Here, we define a program formulation for CMs, and show a conversion method between an RCM in the quadruple form and an RCM in the program form. Note that a similar program formulation for RTMs is proposed in [50].

4.2.3.1 The instruction set and a program for CMs

Instructions for a CM(k) are the following:

$$I_i, D_i, B_i(b_0, b_1), M_i(m_0, m_1), \text{ and } H,$$

where b_0, b_1, m_0 and m_1 are addresses of instructions, and $i \in \{1, \ldots, k\}$. Intuitive meanings of the instructions are as follows.

I_i	Increment the i-th counter
D_i	Decrement the i-th counter
$B_i(b_0, b_1)$	Branch on the contents of the i-th counter, *i.e.*,
	if the i-th counter is 0, then go to b_0, else go to b_1
$M_i(m_0, m_1)$	Merge on the contents of the i-th counter, *i.e.*,
	if the i-th counter is 0, then merge from m_0, else from m_1
H	Halt.

To define a program for a CM(k), we give the sets $A^L, \mathbf{B}_k^L$ and $\mathbf{M}_k^L$ as follows, where L (> 0) is the length of a program.

$$
\begin{aligned}
A^L &= \{0, 1, \ldots, L-1\} \\
\mathbf{B}_k^L &= \{B_i(b_0, b_1) \mid b_0, b_1 \in A^L \cup \{-\}, \ i \in \{1, \ldots, k\}\} \\
\mathbf{M}_k^L &= \{M_i(m_0, m_1) \mid m_0, m_1 \in A^L \cup \{-\}, \ i \in \{1, \ldots, k\}\}
\end{aligned}
$$

Here, A^L is the set of *addresses* of instructions, where the 0-th instruction has the address 0, and the last has $L-1$. The set $\mathbf{B}_k^L$ ($\mathbf{M}_k^L$, respectively) contains all possible $B_i(b_0, b_1)$ instructions ($M_i(m_0, m_1)$ instructions), where $-$ means no address is specified. If $b_p \in A^L$ ($m_p \in A^L$, respectively) for $p \in \{0, 1\}$, it is called a *destination address* (*source address*) of *port p* of the instruction. The set $\mathbf{S}_k^L$ of instructions, which is for a program of length L of CM(k), is as follows.

$$
\mathbf{S}_k^L = \{I_i, D_i \mid i \in \{1, \ldots, k\}\} \cup \mathbf{B}_k^L \cup \mathbf{M}_k^L \cup \{H\}
$$

Definition 4.12. A *well-formed program* (WFP) P of length L for CM(k) is a mapping $P : A^L \to \mathbf{S}_k^L$ that satisfies the following syntactic constraints.

(C1) The last instruction must be H or B_i instruction:

$$
P(L-1) \in \{H\} \cup \mathbf{B}_k^L
$$

(C2) The 0-th instruction must not be an M_i instruction, and the instruction just before M_i must be an H or B_i instruction:

$$
P(0) \notin \mathbf{M}_k^L \ \wedge \ \forall a \in A^L - \{0\}(P(a) \in \mathbf{M}_k^L \Rightarrow P(a-1) \in \{H\} \cup \mathbf{B}_k^L)
$$

(C3) If the instruction of the address a is B_i, and its port p has a destination address a', then the instruction at the address a' must be M_i, and its port p has the source address a:

$$
\begin{aligned}
&\forall a, a' \in A^L, \ \forall p \in \{0, 1\}, \ \forall i \in \{1, \ldots, k\}, \\
&\forall b_0, b_1 \in A^L \cup \{-\}, \ \exists m_0, m_1 \in A^L \cup \{-\} \\
&((P(a) = B_i(b_0, b_1) \wedge b_p = a') \Rightarrow (P(a') = M_i(m_0, m_1) \wedge m_p = a))
\end{aligned}
$$

(C4) If the instruction of the address a is M_i, and its port p has a source address a', then the instruction at the address a' must be B_i, and its port p has the destination address a:

$$\forall a, a' \in A^L, \ \forall p \in \{0,1\}, \ \forall i \in \{1, \ldots, k\},$$
$$\forall m_0, m_1 \in A^L \cup \{-\}, \ \exists b_0, b_1 \in A^L \cup \{-\}$$
$$((P(a) = M_i(m_0, m_1) \wedge m_p = a') \Rightarrow (P(a') = B_i(b_0, b_1) \wedge b_p = a))$$

The constraint (C1) prevents the case of going to the address L. The constraint (C2) guarantees that each M_i instruction is activated only by B_i instructions. The constraints (C3) and (C4) say that the destination addresses of port p of B_i instructions, and the source addresses of port p of M_i instructions have one-to-one correspondence for each $p \in \{0, 1\}$.

Example 4.6. Let P_{move} be the following sequence of instructions.

$$
\begin{array}{ccccccccc}
0 & 1 & 2 & 3 & 4 & 5 & 6 & 7 & 8 \\
B_2(1, -) & M_2(0, 6) & D_1 & I_2 & B_1(7, 5) & M_1(-, 4) & B_2(-, 1) & M_1(4, -) & H
\end{array}
$$

We can verify that P_{move} satisfies the constraints (C1)–(C4) in Definition 4.12. Therefore, it is a well-formed program (WFP) of a CM(2). Since it is cumbersome to follow addresses in B_i and M_i instructions, we often draw a WFP in a graphical form as in Fig. 4.3. How this WFP works will be shown in Example 4.7.

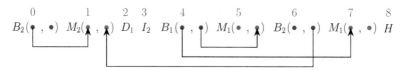

Fig. 4.3   Graphical representation of the WFP P_{move}.

We now define a CM M in the program form, which has a WFP.

Definition 4.13. A *CM in the program form* is defined by

$$M = (P, k, A_F)$$

where P is a WFP of length L, k is the number of counters, and A_F is a set of *final addresses* that satisfy the following: $A_F \subseteq \{a \mid a \in A^L \wedge P(a) = H\}$, where $A^L = \{0, \ldots, L-1\}$.

In Theorem 4.6, it will be shown that any CM of the program form can be converted into an equivalent RCM in the quadruple form. By this, we can see that CMs of the program form are actually *reversible* CMs.

An instantaneous description (ID) of a CM in the program form, and a transition relation among IDs are defined next.

Definition 4.14. Let $M = (P, k, A_F)$ be a CM in the program form. Let L be the length of P. Thus the set of addresses of P is $A^L = \{0, \ldots, L-1\}$. An *instantaneous description* (ID) of M is an expression $(a, (n_1, n_2, \ldots, n_k)) \in A^L \times \mathbb{N}^k$. It represents that the i-th counter keeps n_i ($i \in \{1, \ldots, k\}$), and the instruction $P(a)$ is going to be executed.

Definition 4.15. Let $M = (P, k, A_F)$ be a CM in the program form, and L be the length of P. The *transition relation* $\vdash_M$ over IDs of M is defined as follows. For every $i \in \{1, \ldots, k\}$, $a, a' \in A^L$ and $n_1, \ldots, n_k, n_i' \in \mathbb{N}$,

$$(a, (n_1, \ldots, n_{i-1}, n_i, n_{i+1}, \ldots, n_k)) \vdash_M (a', (n_1, \ldots, n_{i-1}, n_i', n_{i+1}, \ldots, n_k))$$

holds if and only if one of the following conditions $(1) - (5)$ is satisfied.

(1) $P(a) = I_i \ \wedge \ n_i' = n_i + 1 \ \wedge \ a' = a + 1$
(2) $P(a) = D_i \ \wedge \ n_i' = n_i - 1 \geq 0 \ \wedge \ a' = a + 1$
(3) $P(a) = B_i(b_0, b_1) \ \wedge \ n_i' = n_i = 0 \ \wedge \ a' = b_0$
(4) $P(a) = B_i(b_0, b_1) \ \wedge \ n_i' = n_i > 0 \ \wedge \ a' = b_1$
(5) $P(a) = M_i(m_0, m_1) \ \wedge \ n_i' = n_i \ \wedge \ a' = a + 1$

Reflexive and transitive closure of $\vdash_M$ is denoted by $\vdash_M^*$, and n-step transition by $\vdash_M^n$ ($n = 0, 1, \ldots$).

Let $M = (P, k, A_F)$ be a CM. An ID $(a, (n_1, \ldots, n_k))$ is called an *initial ID* of M, if $a = 0$. An ID C is a *halting ID*, if there is no ID C' such that $C \vdash_M C'$. An ID $(a, (n_1, \ldots, n_k))$ of M is called a *final ID*, if $a \in A_F$. Every final ID $(a, (n_1, \ldots, n_k))$ is a halting ID, since $P(a) = H$. Let C_i ($i \in \{0, 1, \ldots, n\}$) be IDs. We say that $C_0 \vdash_M C_1 \vdash_M \cdots \vdash_M C_n$ (or $C_0 \vdash_M^* C_n$) is a *complete computing process* of M, if C_0 is an initial ID, and C_n is a final ID.

Example 4.7. Consider a CM $M_{\text{move}} = (P_{\text{move}}, 2, \{8\})$, where P_{move} is the WFP in Example 4.6. If we start from the initial ID $(0, (2, 0))$, we have the following complete computing process of M_{move}.

$$(0, (2, 0)) \vdash_{M_{\text{move}}} (1, (2, 0)) \vdash_{M_{\text{move}}} (2, (2, 0)) \vdash_{M_{\text{move}}} (3, (1, 0)) \vdash_{M_{\text{move}}}^9 (8, (0, 2))$$

Generally, $(0, (x, 0)) \vdash_{M_{\text{move}}}^* (8, (0, x))$ holds for all $x > 0$, *i.e.*, M_{move} moves the contents of the counter 1 to the counter 2.

4.2.4   *Conversion between the quadruple form and the program form*

We show that any CM in the program form can be converted into an RCM in the quadruple form that simulates the former. Conversely, any RCM in the quadruple form can be converted into a CM in the program form that simulates the former. Hence, the class of CMs in the program form exactly characterizes the class of RCMs in the quadruple form.

Theorem 4.6. *Let* $M = (P, k, A_F)$ *be a CM in the program form, and let L be the length of P. Then, we can construct an RCM $M' = (\{q_0, \ldots, q_{L-1}\}, k, \delta, q_0, \{q_j \mid j \in A_F\})$ in the quadruple form that simulates M in the following way, where $A^L = \{0, \ldots, L-1\}$.*

$$\forall m \in \mathbb{N}, \forall a \in A^L, \forall n_1, \ldots, n_k, n'_1, \ldots, n'_k \in \mathbb{N}$$
$$(0, (n_1, \ldots, n_k)) \;\vdash^{\frac{m}{M}}\; (a, (n'_1, \ldots, n'_k))$$
$$\Leftrightarrow (q_0, (n_1, \ldots, n_k)) \;\vdash^{\frac{m}{M'}}\; (q_a, (n'_1, \ldots, n'_k))$$

Proof. The set δ of quadruples is obtained by executing $(1)-(5)$ for each $a \in A^L$, where $i \in \{1, \ldots, k\}$ and $b_0, b_1, m_0, m_1 \in A^L \cup \{-\}$.

(1) If $P(a) = I_i$, include $[q_a, i, +, q_{a+1}]$ in δ.
(2) If $P(a) = D_i$, include $[q_a, i, -, q_{a+1}]$ in δ.
(3) If $P(a) = B_i(b_0, b_1)$, then
     include $[q_a, i, Z, q_{b_0}]$ in δ if $b_0 \in A^L$, and
     include $[q_a, i, P, q_{b_1}]$ in δ if $b_1 \in A^L$.
(4) If $P(a) = M_i(m_0, m_1)$, include $[q_a, i, 0, q_{a+1}]$ in δ.
(5) If $P(a) = H$, include nothing in δ.

Since each state q_a directly simulates the instruction $P(a)$ by the above quadruples, it is easy to show that M' correctly simulates M as described in the theorem by a mathematical induction on the number m of steps (hence its proof is omitted).

We now show M' is reversible. First, we can see that the states q_{a+1} in (1), (2) or (4) appears nowhere else as the fourth component of a quadruple. In particular, q_{b_j} (when $b_j \neq -$) in (3) cannot be the same as q_{a+1} for each $j \in \{0, 1\}$. This is because $P(b_j)$ must be an M_i instruction, and hence $P(b_j - 1)$ is neither I, D nor M instruction. Therefore, quadruples that are added in (1), (2) and (4) do not violate the reversibility condition.

Next, consider the procedure (3) above. Since $P(a) = B_i(b_0, b_1)$, we can see $P(b_0) = M_i(a, m_1)$ (if $b_0 \neq -$) and $P(b_1) = M_i(m_0, a)$ (if $b_1 \neq -$) for some $m_0, m_1 \in A^L \cup \{-\}$. Consider the case $P(b_0) = M_i(a, m_1)$. In this case $[q_a, i, Z, q_{b_0}]$ is included in δ by (3). If $m_1 \neq -$, then $P(m_1) = B_i(b'_0, b_1)$

for some $b_0' \in A^L \cup \{-\}$, and thus $[q_{m_1}, i, P, q_{b_0}]$ is included in δ. There is no other quadruple that has q_{b_0} as the fourth component since $P(b_0) = M_i(a, m_1)$. Apparently, the pair $[q_a, i, Z, q_{b_0}]$ and $[q_{m_1}, i, P, q_{b_0}]$ satisfies the reversibility condition. On the other hand, if $m_1 = -$, then $[q_a, i, Z, q_{b_0}]$ is the only quadruple that has q_{b_0} as the fourth quadruple, and, of course, it satisfies the reversibility condition. The case $P(b_1) = M_i(m_0, a)$ is similar to the above case.

By above, we can conclude M' is a reversible CM. □

Example 4.8. Let $M_{\text{move}} = (P_{\text{move}}, 2, \{8\})$ be a CM in the program form in Example 4.7. The WFP P_{move} is as follows, which is shown in Example 4.6.

$$\begin{array}{ccccccccc} 0 & 1 & 2 & 3 & 4 & 5 & 6 & 7 & 8 \\ B_2(1,-) & M_2(0,6) & D_1 & I_2 & B_1(7,5) & M_1(-,4) & B_2(-,1) & M_1(4,-) & H \end{array}$$

By the method shown in Theorem 4.6, we have an RCM in the quadruple form. It is $M'_{\text{move}} = (\{q_0, \ldots, q_8\}, 2, \delta_{\text{move}}, q_0, \{q_8\})$, where δ_{move} is:

$$\delta_{\text{move}} = \{ \ [q_0, 2, Z, q_1], \ [q_1, 2, 0, q_2], \ [q_2, 1, -, q_3], \ [q_3, 2, +, q_4],$$
$$[q_4, 1, Z, q_7], \ [q_4, 1, P, q_5], \ [q_5, 1, 0, q_6], \ [q_6, 2, P, q_1], \ [q_7, 1, 0, q_8] \ \}$$

Note that the quadruples $[q_1, 2, 0, q_2]$, $[q_5, 1, 0, q_6]$, and $[q_7, 1, 0, q_8]$ perform no-operation on the counters. Therefore, we can simplify M'_{move} by removing these quadruples and the states q_1, q_5 and q_7. The resulting RCM is $M''_{\text{move}} = (\{q_0, q_2, q_3, q_4, q_6, q_8\}, 2, \hat{\delta}_{\text{move}}, q_0, \{q_8\})$ where $\hat{\delta}_{\text{move}}$ is:

$$\hat{\delta}_{\text{move}} = \{ \ [q_0, 2, Z, q_2], \ [q_2, 1, -, q_3], \ [q_3, 2, +, q_4],$$
$$[q_4, 1, Z, q_8], \ [q_4, 1, P, q_6], \ [q_6, 2, P, q_1] \ \}$$

Next, we show that any RCM(k) M in the quadruple form can be converted into an RCM(k) M' in the program form. We first prepare a *dummy instruction* "N," which performs no-operation on the counters, and add it to the set of instructions $\mathbf{S}_k^L$. It is for making it easy to calculate source and destination addresses of M_i and B_i instructions. After a program P over the set $\mathbf{S}_k^L \cup \{N\}$ that simulates M is obtained, we simplify P by removing occurrences of N and other useless instructions.

Theorem 4.7. *Let* $M = (\{q_0, \ldots, q_{n-1}\}, k, \delta, q_0, \{q_{n-1}\})$ *be an RCM in the quadruple form. We assume* M *is in the normal form (Definition 4.11). Then, we can construct an RCM* $M' = (P, k, \{3n-1\})$ *in the program form over the instruction set* $\mathbf{S}_k^{3n} \cup \{N\}$ *that simulates* M *in the following way.*

$$\forall m \in \mathbb{N}, \forall h \in \{0, \ldots, n-1\}, \forall n_1, \ldots, n_k, n_1', \ldots, n_k' \in \mathbb{N}$$
$$(q_0, (n_1, \ldots, n_k)) \ \vdash^{\frac{m}{M}} \ (q_h, (n_1', \ldots, n_k'))$$
$$\Leftrightarrow (0, (n_1, \ldots, n_k)) \ \vdash^{\frac{3m}{M'}} \ (3h, (n_1', \ldots, n_k'))$$

Furthermore, the following holds for the final state q_{n-1}.

$$\forall m \in \mathbb{N}, \forall n_1, \ldots, n_k, n'_1, \ldots, n'_k \in \mathbb{N}$$
$$(q_0, (n_1, \ldots, n_k)) \models^{\frac{m}{M}} (q_{n-1}, (n'_1, \ldots, n'_k))$$
$$\Leftrightarrow (0, (n_1, \ldots, n_k)) \models^{\frac{3m}{M'}+2} (3n - 1, (n'_1, \ldots, n'_k))$$

Proof. The program P of M' is of length $3n$. The operation performed by the state q_s of M is simulated by the three consecutive instructions $P(3s)$, $P(3s + 1)$ and $P(3s + 2)$ defined below. Note that in the following formulas, we assume $i \in \{1, \ldots, k\}$, $r, r_0, r_1 \in \{0, \ldots, n - 2\}$, and $t, t_0, t_1 \in \{1, \ldots, n - 1\}$.

$$P(3s) = \begin{cases} M_i(3r_0 + 2, 3r_1 + 2) & \text{if } \exists i\,(\exists r_0, r_1([q_{r_0}, i, Z, q_s], [q_{r_1}, i, P, q_s] \in \delta)) \\ M_i(3r_0 + 2, -) & \text{if } \exists i\,(\exists r_0([q_{r_0}, i, Z, q_s] \in \delta) \\ & \quad \wedge \forall r_1([q_{r_1}, i, P, q_s] \notin \delta)) \\ M_i(-, 3r_1 + 2) & \text{if } \exists i\,(\exists r_1([q_{r_1}, i, P, q_s] \in \delta) \\ & \quad \wedge \forall r_0([q_{r_0}, i, Z, q_s] \notin \delta)) \\ M_i(3r + 2, 3r + 2) & \text{if } \exists i\,\exists r\,\exists d \in \{+, -\}([q_r, i, d, q_s] \in \delta) \\ N & \text{elsewhere } (\textit{i.e.}, q_s \text{ has no previous state}) \end{cases}$$

$$P(3s + 1) = \begin{cases} I_i & \text{if } \exists i\,\exists t\,([q_s, i, +, q_t] \in \delta) \\ D_i & \text{if } \exists i\,\exists t\,([q_s, i, -, q_t] \in \delta) \\ N & \text{elsewhere} \end{cases}$$

$$P(3s + 2) = \begin{cases} B_i(3t_0, 3t_1) & \text{if } \exists i\,(\exists t_0, t_1([q_s, i, Z, q_{t_0}], [q_s, i, P, q_{t_1}] \in \delta)) \\ B_i(3t_0, -) & \text{if } \exists i\,(\exists t_0([q_s, i, Z, q_{t_0}] \in \delta) \\ & \quad \wedge \forall t_1([q_s, i, P, q_{t_1}] \notin \delta)) \\ B_i(-, 3t_1) & \text{if } \exists i\,(\exists t_1([q_s, i, P, q_{t_0}] \in \delta) \\ & \quad \wedge \forall t_0([q_s, i, Z, q_{t_0}] \notin \delta)) \\ B_i(3t, 3t) & \text{if } \exists i\,\exists t\,\exists d \in \{+, -\}([q_s, i, d, q_t] \in \delta) \\ H & \text{elsewhere } (\textit{i.e.}, q_s \text{ is a halting state}) \end{cases}$$

It is easy to see that the values of $P(3s)$, $P(3s + 1)$ and $P(3s + 2)$ are uniquely determined by the above formulas, since M is deterministic and reversible. We can also see that P satisfies the constraints (C1)–(C4) in Definition 4.12. Therefore, it is a WFP, and thus M' is an RCM(k) in the program form.

We now verify that the three consecutive instructions correctly simulates the operation performed by q_s. First, consider the instruction $P(3s)$. If $q_s = q_0$, then it has no previous state, since M is in the normal form. In this case, $P(3s) = P(0) = N$. On the other hand, if there is a quadruple of the form $[q_r, i, x, q_s]$, then q_s may come from q_r. When $[q_{r_0}, i, Z, q_s]$ and

$[q_{r_1}, i, P, q_s]$ exist, two different execution paths are merged into one by the instruction $P(3s) = M_i(3r_0 + 2, 3r_1 + 2)$. If only $[q_{r_0}, i, Z, q_s]$ ($[q_{r_1}, i, P, q_s]$, respectively) exists, the instruction $P(3s) = M_i(3r_0+2, -)$ ($M_i(-, 3r_1+2)$) connects the source address $3r_0 + 2$ ($3r_1 + 2$) and the destination address $3s$. On the other hand, when $[q_r, i, d, q_s]$ ($d \in \{+, -\}$) exists, $P(3s) = M_i(3r + 2, 3r + 2)$ connects the source address $3r + 2$ and the destination address $3s$, which actually works an unconditional jump.

Second, consider the instruction $P(3s + 1)$. When $[q_s, i, +, q_t]$ ($[q_r, i, -, q_s]$, respectively) exists, the increment (decrement) operation is directly simulated by $P(3s + 1) = I_i$ (D_i). In the case where there is a quadruple of the form $[q_s, i, x, q_t]$ ($x \in \{Z, P\}$) or q_s is a halting state, $P(3s + 1) = N$.

Third, consider the instruction $P(3s + 2)$. When both $[q_s, i, Z, q_{t_0}]$ and $[q_s, i, P, q_{t_1}]$ exist, branching of the execution path is realized by the instruction $P(3s+2) = B_i(3t_0, 3t_1)$. If only $[q_s, i, Z, q_{t_0}]$ ($[q_s, i, P, q_{t_1}]$, respectively) exists, jumping to q_{t_0} (q_{t_1}) is realized by $P(3s+2) = B_i(3t_0, -)$ ($B_i(-, 3t_1)$). When $[q_s, i, d, q_t]$ ($d \in \{+, -\}$) exists, unconditional jump to q_t is realized by $P(3s+2) = B_i(3t, 3t)$. If q_s is a halting state, M' halts by $P(3s+2) = H$.

By the above observation, we can see that each step of M is correctly simulated in three steps of M'. Note that, more precisely, the theorem is proved by a mathematical induction on the number m of steps of M, though we omit it here. $\square$

The following example shows how the conversion method in Theorem 4.7 is applied. The program constructed by this method contains many occurrences of N. It also contains useless sequences of instructions. Simplification of the program is also shown in the example.

Example 4.9. Consider the RCM $M_{\text{twice}} = (\{q_0, \ldots, q_6\}, 2, \delta_{\text{twice}}, q_0, \{q_6\})$ in Example 4.4, where δ_{twice} is as follows.

$$\delta_{\text{twice}} = \{[q_0, 2, Z, q_1], [q_1, 1, Z, q_6], [q_1, 1, P, q_2], [q_2, 1, -, q_3],$$
$$[q_3, 2, +, q_4], [q_4, 2, +, q_5], [q_5, 2, P, q_1]\}$$

M_{twice} computes $g(x) = 2x$. For example, $(q_0, (2, 0)) \mid\frac{12}{M_{\text{twice}}} (q_6, (0, 4))$ when $x = 2$. The WFP P_{twice} of the RCM $M'_{\text{twice}} = (P_{\text{twice}}, 2, \{20\})$ constructed by the method of Theorem 4.7 is shown in Fig. 4.4. It simulates the above complete computing process by $(0, (2, 0)) \mid\frac{38}{M'_{\text{twice}}} (20, (0, 4))$.

However, P_{twice} has many useless instructions. First, we remove each occurrence of the instruction N. When an occurrence of N at the address

a is removed, all the source/destination addresses of M_i and B_i instructions, which are greater than a, must be decreased by 1. This procedure of removing N is repeated until no N exists.

Second, if there are consecutive two instructions $B_i(a+1, a+1)$ and $M_i(a, a)$ at the addresses a and $a+1$ for some $i \in \{1, \ldots, k\}$, then we remove them. Actually, an unconditional jump becomes meaningless when it jumps to the next address. After removing them, all the source/destination addresses of M_i and B_i instructions, which are greater than a, must be decreased by 2. This procedure is also repeated until no such pair exists. In P_{twice}, there are three occurrences of such instructions at the address pairs $(8, 9), (11, 12)$ and $(14, 15)$.

The program $\hat{P}_{\text{twice}}$ shown in Fig. 4.5 is the simplified WFP obtained from P_{twice} by the above method. Thus, we have $\hat{M}_{\text{twice}} = (\hat{P}_{\text{twice}}, 2, \{9\})$ that simulates M_{twice}. The complete computing process of $\hat{M}_{\text{twice}}$ for the input $x = 2$ is as follows: $(0, (2, 0)) \vdash_{\hat{M}_{\text{twice}}}^{18} (9, (0, 4))$.

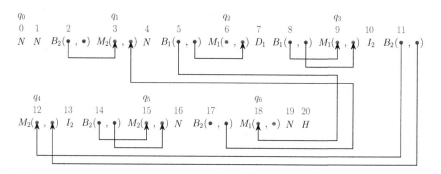

Fig. 4.4   The WFP P_{twice} converted from M_{twice} by the method of Theorem 4.7.

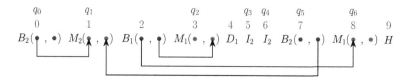

Fig. 4.5   The simplified WFP $\hat{P}_{\text{twice}}$ for $\hat{M}_{\text{twice}}$.

Computing processes of the RCMs in the program form given in Examples 4.6 and 4.9 are observed by the simulator contained in [21].

4.3   Reversible Logic Element with Memory (RLEM)

A *reversible logic element with memory* (RLEM), which was first introduced in [51], is an important logic element for implementing universal reversible computing systems in a reversible CA. This is because RTMs and RCMs can be constructed out of RLEMs more easily than to use reversible logic gates, as it will be explained in Sec. 4.3.7. An RLEM is a kind of a *reversible finite automaton* having output symbols as well as input symbols, which is called a reversible sequential machine of Mealy type.

4.3.1   *Definition of RLEM*

Definition 4.16. A *sequential machine* (SM) is defined by $M = (Q, \Sigma, \Gamma, \delta)$, where Q is a finite set of states, Σ and Γ are finite sets of input and output symbols, and $\delta : Q \times \Sigma \to Q \times \Gamma$ is a move function (Fig. 4.6(a)). If δ is injective, it is called a *reversible sequential machine* (RSM).

Definition 4.17. A *reversible logic element with memory* (RLEM) is an RSM $M = (Q, \Sigma, \Gamma, \delta)$ that satisfies $|\Sigma| = |\Gamma|$. M is called a $|Q|$-state $|\Sigma|$-symbol RLEM.

To use an SM as a logic element, we interpret it as the one with decoded input ports and output ports (Fig. 4.6(b)). Namely, for each input symbol, there is a unique input port to which a signal (or a particle) is given. It is also the case for the output symbols. Therefore, signals should not be given to two or more input ports at the same time.

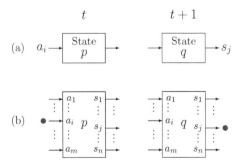

Fig. 4.6   (a) A sequential machine (SM) with $\delta(p, a_i) = (q, s_j)$, and (b) an interpretation of it as a module having decoded input ports and output ports.

RLEM-circuit. When connecting many RLEMs to compose a circuit, each output port of an RLEM can be connected to at most one input port of another (or may be the same) RLEM. Furthermore, two or more output ports should not be connected to one input port. Therefore, neither branching (*i.e.*, fan-out of an output) nor merging of signal lines is permitted. See Sect. 3.5.1 of [9] for the precise definition of an RLEM-circuit.

4.3.2 *Rotary element (RE), a typical 2-state RLEM*

Among RLEMs, 2-state RLEMs are particularly important, since they are simple yet powerful (see Sec. 4.3.4). We first consider a typical 2-state RLEM called a rotary element.

A *rotary element* (RE) [51] is a 2-state RLEM that has four input ports and four output ports, and is depicted as in Fig. 4.7. Intuitively, an RE has a rotatable bar inside, and an incoming signal is controlled by the bar. It takes either of the two states V or H, depending on the direction of the bar. If the direction of a coming signal is parallel to the bar, the signal goes straight ahead, and the state does not change (Fig. 4.8(a)). If the direction of a coming signal is orthogonal to the bar, the signal turns right, and the state changes (Fig. 4.8(b)). An RE is formally defined by an RSM $M_{RE} = (\{V, H\}, \{n, e, s, w\}, \{n', e', s', w'\}, \delta_{RE})$, where δ_{RE} is given in Table 4.1.

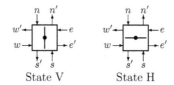

State V   State H

Fig. 4.7   Two states V and H of a rotary element (RE).

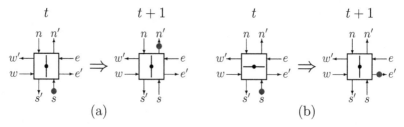

Fig. 4.8   Operations of an RE. (a) Parallel case, and (b) orthogonal case.

Table 4.1　Move function δ_{RE} of an RE.

Present state	Input			
	n	e	s	w
V	V s'	H n'	V n'	H s'
H	V w'	H w'	V e'	H e'

In the following, we often use REs to compose RTMs. This is because the operation of RE is intuitively easy to understand, and RTMs can be constructed by it very simply (Sec. 4.3.5).

We now consider the *billiard-ball model* (BBM) proposed by Fredkin and Toffoli [52]. It is an idealized mechanical model of computing where moving balls collide with other balls or reflectors. It is assumed that collisions are elastic, and there is no friction. Fredkin and Toffoli showed that a *Fredkin gate*, a universal reversible logic gate, can be realized in BBM.

Though an RE can be constructed out of Fredkin gates and delay elements, there is a direct and compact implementation method of an RE in BBM [9]. A configuration of BBM that simulates an RE is given in Fig. 4.9. In this figure, a stationary ball called a *state-ball* is put at the position H or V, which has integer coordinates. The diameter of a ball is assumed to be $\sqrt{2}$. Small rectangles in the figure are reflectors. Note that the size of the reflectors are set to be very small in order to reduce the size of the BBM configuration, but it is not essential. A moving ball called a *signal-ball* can be given to any one of the input ports n, e, s and w. It may be reflected by some of the reflectors, and may interact with the state-ball. In Fig. 4.9, the trajectories of balls for all the combinations of the state and the input are written in one figure, and thus it may look complicated. However, the key idea is simple as explained below. We consider only the case where the state is V and the input is s, and the case where the state is H and the input is s. Other cases are similar.

Figure 4.10 shows the case where the RE is in the state V and a signal-ball is given to s. In this case, the signal-ball simply goes straight ahead interacting neither with the state-ball nor with reflectors. It finally comes out from the output port n'. Thus, $\delta_{\mathrm{RE}}(V, s) = (V, n')$ is simulated (see also Fig. 4.8(a)).

Figure 4.11 shows the case where the RE is in the state H and a signal-ball is given to s. First, the signal-ball collides with the state-ball. Then, the state-ball and the signal-ball travel along the paths labeled by s_0 and

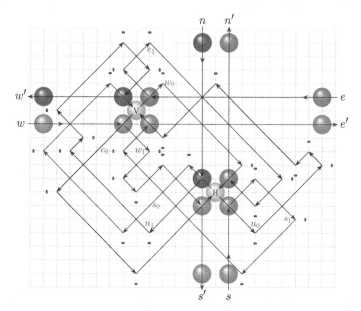

Fig. 4.9   Rotary element (RE) realized in the billiard-ball model (BBM) [9].

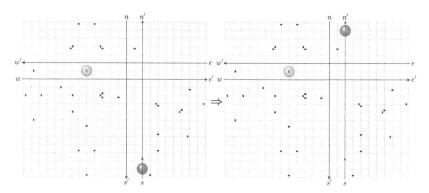

Fig. 4.10   Process of simulating $\delta_{\mathrm{RE}}(V, s) = (V, n')$ of an RE in BBM.

s_1, respectively. Next, the balls collide again. By the second collision, the state-ball stops at the position V, while the signal-ball continues to travel rightward, and finally goes out from the port e'. Note that since the second collision is a reverse process of the first collision, the state-ball stops at V. By this, $\delta_{\mathrm{RE}}(H, s) = (V, e')$ is simulated (Fig. 4.8(b)).

An advantage of using this method is that, in an idle state, the state-ball is stationary. Therefore, a signal-ball can be given at any moment and with

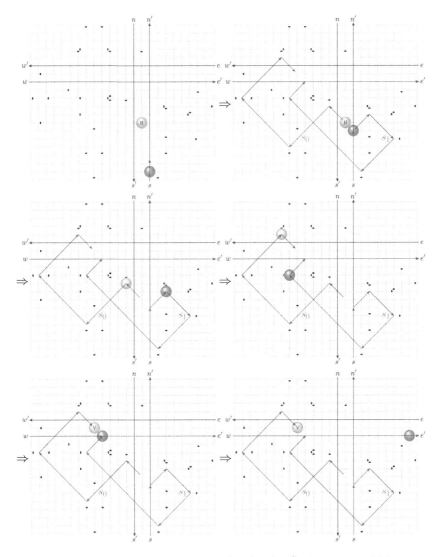

Fig. 4.11   Process of simulating $\delta_{\mathrm{RE}}(H, s) = (V, e')$ of an RE in BBM.

any speed. Hence, there is no need to synchronize the timing of the two balls. On the other hand, if we use a moving ball to memorize the state, then some synchronization mechanism is necessary so that two moving balls interact correctly. An implementation method for other RLEMs in BBM is given in [9, 53].

In Sec. 6.1.4, the above idea, which uses a stationary ball to keep the state of an RE, will be applied to implement it in a simple reversible CA. There, a special small pattern of the CA called a position marker is used to keep the state, and a state-change is performed by shifting the position marker by colliding a signal.

4.3.3   *2-State RLEMs and their classification*

In this section, we investigate the class of 2-state RLEMs. We first give an *identification number* to each of 2-state k-symbol RLEMs. Let $M = (Q, \Sigma, \Gamma, \delta)$ be an RLEM such that $|Q| = 2$ and $|\Sigma| = |\Gamma| = k$. Since δ is a bijection, there are $(2k)!$ kinds of δ's. Here, we fix the sets Q, Σ and Γ as follows: $Q = \{0, 1\}$, $\Sigma = \{a_1, \ldots, a_k\}$ and $\Gamma = \{s_1, \ldots, s_k\}$. Thus, a bijection δ can be specified by a permutation on the set $\{0, 1\} \times \{s_1, \ldots, s_k\}$. All the move functions δ's of 2-state k-symbol RLEMs are numbered by $0, \ldots, (2k)! - 1$ in the lexicographic order of permutations. An identification number is obtained by attaching the prefix "k-" to the serial number, *e.g.*, 4-289. The RLEM with an identification number k-n is denoted by RLEM k-n.

Next, we show a method of representing a move function of a 2-state RLEM graphically. It makes it easy to describe a move function. Here, the method is explained using the following example (its detailed definition is given in [9]).

Example 4.10. Consider RLEM 4-289: $M_{4\text{-}289} = (\{0, 1\}, \{a_1, a_2, a_3, a_4\}, \{s_1, s_2, s_3, s_4\}, \delta_{4\text{-}289})$. Its move function $\delta_{4\text{-}289}$ is described in Table 4.2. Figure 4.12 shows a graphical representation of $\delta_{4\text{-}289}$. In this figure, each of the two states are represented by a rectangle having input ports and output ports. The relation between the inputs and the outputs is indicated by solid and dotted lines. We assume a signal (or a particle) is given to at most one input port at a time. If a particle is given to an input port, it moves along the line connected to the input port, and goes out from the output port connected to it. If a particle goes through a dotted line, the state does not change (Fig. 4.13(a)). On the other hand, if it goes through a solid line, the state changes to the other (Fig. 4.13(b)).

There are $(2k)!$ kinds of 2-state k-symbol RLEMs. However, among them there are many equivalent RLEMs, which are obtained by renaming the states and/or the input/output symbols. Hence, the total number of essentially different 2-state k-symbol RLEMs decreases significantly. Now, equivalence among RLEMs are defined as follows.

Table 4.2   Move function $\delta_{4\text{-}289}$.

Present state	Input			
	a_1	a_2	a_3	a_4
0	0 s_1	0 s_2	1 s_1	1 s_2
1	0 s_3	0 s_4	1 s_4	1 s_3

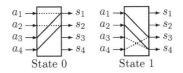

Fig. 4.12   Graphical representation of RLEM 4-289.

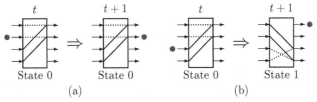

Fig. 4.13   (a) If a particle passes a dotted line, the state remains to be the same. (b) If a particle passes a solid line, the state changes to the other.

Definition 4.18. Let $M_1 = (Q_1, \Sigma_1, \Gamma_1, \delta_1)$ and $M_2 = (Q_2, \Sigma_2, \Gamma_2, \delta_2)$ be two RLEMs such that $|Q_1| = |Q_2|$, $|\Sigma_1| = |\Sigma_2| = |\Gamma_1| = |\Gamma_2|$. The RLEMs M_1 and M_2 are called *equivalent* and denoted by $M_1 \sim M_2$, if there exist bijections $f : Q_1 \to Q_2$, $g : \Sigma_1 \to \Sigma_2$, and $h : \Gamma_1 \to \Gamma_2$ that satisfy

$$\forall q \in Q_1, \ \forall a \in \Sigma_1 \ (\delta_1(q, a) = \psi(\delta_2(f(q), g(a))))$$

where $\psi : Q_2 \times \Gamma_2 \to Q_1 \times \Gamma_1$ is defined as follows.

$$\forall q \in Q_2, \ \forall s \in \Gamma_2 \ (\psi(q, s) = (f^{-1}(q), h^{-1}(s)))$$

It is easy to see that the relation "$\sim$" is an equivalence relation, *i.e.*, it is reflexive, symmetric, and transitive.

Example 4.11. An RE (*i.e.*, M_{RE} in Sec. 4.3.2) and RLEM 4-289 in Example 4.10 are equivalent under the relation $\sim$. It is shown by using the following bijections f, g and h.

$$f(\mathrm{V}) = 0, \ \ f(\mathrm{H}) = 1$$
$$g(n) = a_1, \ g(e) = a_4, \ g(s) = a_2, \ g(w) = a_3$$
$$h(n') = s_2, \ h(e') = s_4, \ h(s') = s_1, \ h(w') = s_3$$

For example, the following holds.

$$\begin{aligned}
\delta_{\mathrm{RE}}(\mathrm{H}, s) &= \psi(\delta_{4\text{-}289}(f(\mathrm{H}), g(s))) \\
&= \psi(\delta_{4\text{-}289}(1, a_2)) \\
&= \psi(0, s_4) \\
&= (f^{-1}(0), h^{-1}(s_4)) \\
&= (\mathrm{V}, e')
\end{aligned}$$

Other cases are also verified similarly. Hence, RE $\sim$ RLEM 4-289.

The total numbers of different 2-state 2-, 3- and 4-symbol RLEMs are 24, 720 and 40320, respectively. On the other hand, the numbers of equivalence classes of 2-, 3- and 4-symbol RLEMs are 8, 24 and 82, respectively [9,54]. Figure 4.14 shows all representative RLEMs in the equivalence classes of 2- and 3-symbol RLEMs. Each representatives are chosen so that it has the smallest identification number in the class.

Among RLEMs, there are some *degenerate* ones that are further equivalent to connecting wires, or to a simpler RLEM with fewer symbols. A more precise definition of degeneracy is found in [9]. In Fig. 4.14, degenerate ones are indicated at the upper-right corner of a box. For example, RLEMs 3-1, 3-6 and 3-21 are degenerate ones. RLEM 3-1 is equivalent to three connecting wires, since no state-change can occur. RLEM 3-6 is equivalent to RLEM 2-2 and one connecting wire. RLEM 3-21 is equivalent to three connecting wires, since two states have exactly the same input-output relation. Table 4.3 shows the classification result. Among all RLEMs, non-degenerate ones are the main concern of the study.

Table 4.3   Numbers of equivalence classes of degenerate and non-degenerate k-symbol RLEMs ($k = 2, 3, 4$) [9].

k	All equivalence classes	Degenerate ones	Non-degenerate ones
2	8	4	4
3	24	10	14
4	82	27	55

4.3.4 *Intrinsic universality of RLEM*

There are infinitely many RLEMs even if we consider only 2-state ones. Some RLEMs among them may be very powerful for composing functional modules from them, while some others may be less powerful. Here, we

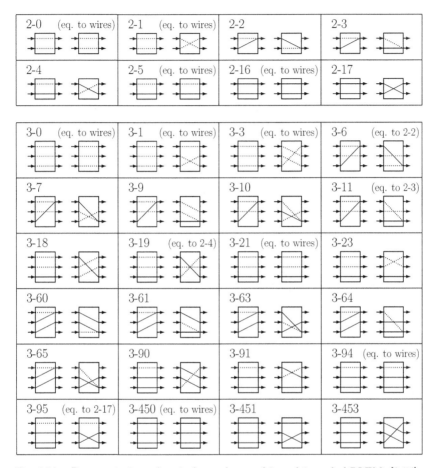

Fig. 4.14   Representatives of equivalence classes of 2- and 3-symbol RLEMs [9, 54].

define the notion of intrinsic universality of an RLEM, and investigate which RLEM is intrinsically universal. It is the property that every RSM can be realized by a circuit composed of the RLEM. As we shall see, it is a remarkable fact that *all* the non-degenerate 2-state RLEM is intrinsically universal except only three 2-symbol RLEMs.

Note that the notion of intrinsic universality comes from the theory of cellular automata (see Sec. 4.5). Further note that, in the past literature, *e.g.*, [9, 54, 55], intrinsic universality was simply written as universality. The reason we use this terminology here is that, in Sec. 4.3.6, we introduce another kind of universality called Turing universality of an RLEM.

Definition 4.19. An RLEM R is called *intrinsically universal* if any RSM M can be realized by a circuit composed only of R.

Note that, here, we do not give a precise definition on the notion of *realization* (see Sec. 3.5.1 of [9] for the details), but it is intuitively clear from the constructed circuits given below. In the following, we say that an RSM M is *realizable* by an RLEM R, if M is simulated by a circuit, say C, composed of R, which is in the form of Fig. 4.6(b). Hence, any circuit composed only of M is realizable by a circuit composed of R by replacing all the occurrences of M by C.

4.3.4.1 *Intrinsic universality of RE*

We first show that any RSM is realizable by an RE. Therefore, an RE is intrinsically universal. To prove it, we introduce a circuit module called an RE-column. RSMs are composed of it systematically.

An *RE-column* of degree n is shown in Fig. 4.15, which has $n + 1$ REs. We assume, in a resting state, it is in the state (a) or (b) of Fig. 4.15, where all the REs except the bottom are in the state V. It has $2n$ input ports $a_1, \ldots, a_n, b_1, \ldots, b_n$, and $2n$ output ports $s_1, \ldots, s_n, t_1, \ldots, t_n$. If a signal is given to one of the input ports, the module will take a state other than those of (a) and (b). However, as we shall see, the module will become again the state (a) or (b) when the signal goes out from it. Therefore, an RE-column behaves as if it is a 2-state RSM. That is to say, the states (a) and (b) are macroscopic states 0 and 1 of the RE-column.

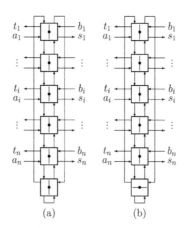

Fig. 4.15   RE-column of degree n. (a) State 0, and (b) state 1.

Table 4.4　The move function of
an RE-column of degree n.

	Input	
Present state	a_i	b_i
0	$0\,s_i$	$1\,s_i$
1	$0\,t_i$	$1\,t_i$

The move function of the RE-column as a 2-state RSM is shown in Table 4.4. We examine how the circuit works for the four cases of state-input pairs: $(0, a_i)$, $(1, a_i)$, $(0, b_i)$, and $(1, b_i)$, where $i \in \{1, \dots, n\}$.

First, consider the case where the state is 0 (Fig. 4.15(a)) and a signal is given to a_i. By the signal from a_i, the i-th RE changes its state from V to H. Then the signal moves downward through the $(n - i + 1)$ REs. At the bottom of the column the signal makes a U-turn, and goes upward through the $(n - i + 1)$ REs. At the i-th RE, the signal turns right and changes the RE's state from H to V. Finally the signal goes out from the port s_i. In this case the RE-column keeps the state 0.

Second, consider the case where the state is 1 (Fig. 4.15(b)) and a signal is given to a_i. As in the first case, the signal sets the i-th RE to the state H, and then moves downward. At the bottom RE, the signal makes a right-turn, and changes the state of the RE to V. The signal goes upward along the left vertical line, and reaches the north input of the top RE. It moves downward through the $(i - 1)$ REs, and makes a right-turn at the i-th RE, restoring the RE's state to V. Finally the signal goes out from the port t_i. In this case the RE-column changes the state from 1 to 0.

Third, consider the case where the state is 0 and a signal is given to b_i. The signal sets the i-th RE to the state H, and moves upward through the $(i - 1)$ REs. Then the signal goes downward along the right vertical line, and reaches the east input of the bottom RE. It changes the state of the bottom RE to H, and moves upward through the $(n - i)$ REs. The signal makes a right-turn at the i-th RE, and restores its state to V. Finally the signal goes out from s_i. The RE-column changes the state from 0 to 1.

Fourth, consider the case where the state is 1 and a signal is given to b_i. As in the third case, the signal sets the i-th RE to the state H, and reaches the east input of the bottom RE. The signal goes out from the west output of the bottom RE without changing its state. It moves upward along the left vertical line, and reaches the north input of the top RE. It makes a right-turn at the i-th RE, restoring the RE's state to V. Finally the signal goes out from t_i. In this case the RE-column keeps the state 1.

We can systematically compose any RSM out of RE-columns. The composing method is explained by the following example of an RSM M_0.

$$M_0 = (\{q_1, q_2, q_3\}, \{c_1, c_2, c_3\}, \{d_1, d_2, d_3\}, \delta_0)$$

The move function δ_0 is given in Table 4.5.

Table 4.5   The move function δ_0 of M_0.

	Present state		
Input	q_1	q_2	q_3
c_1	$q_2\, d_2$	$q_3\, d_2$	$q_3\, d_3$
c_2	$q_3\, d_1$	$q_2\, d_1$	$q_2\, d_3$
c_3	$q_1\, d_2$	$q_1\, d_3$	$q_1\, d_1$

Figure 4.16 shows the circuit that simulates M_0. It consists of three RE-columns of degree 3. The j-th RE-column corresponds to the j-th state q_j of M_0. The i-th row except the bottom row corresponds to the input symbol c_i and the output symbol d_i. If the state of M_0 is q_j, then the state of the j-th RE-column is set to 1, while the other RE-columns are set to 0. Figure 4.16 shows that M_0 is in the state q_3.

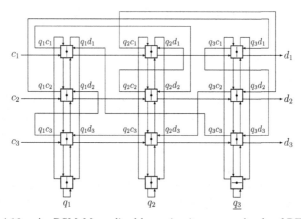

Fig. 4.16   An RSM M_0 realized by a circuit composed only of RE [9].

For example, assume an input signal is given to the port c_2. Since the first two RE-columns are in the state 0, the signal goes rightward through these RE-columns without changing their states. At the third RE-column, the signal changes the RE-column's state from 1 to 0, and then comes out from the west output port of the second RE, which is labeled by q_3c_2. This port is connected to the east input port of the third RE of the second

RE-column labeled by $q_2 d_3$. By this, the state of the second RE-column changes from 0 to 1. The signal appears from the east output port of the third RE in the second RE-column. Since the third RE-column is now in the state 0, the signal finally goes out from d_3. By above, the operation $\delta_0(q_3, c_2) = (q_2, d_3)$ is simulated. Other cases are similar to this case.

In [21], a simulator for RLEM circuits built on Golly is given. There, we can observe how the circuit of Fig. 4.16 works.

Generally, for any RSM $M = (\{q_1, \ldots, q_n\}, \{c_1, \ldots, c_l\}, \{d_1, \ldots, d_m\}, \delta)$, we can construct a circuit composed of RE-columns that simulates M in the following way. First prepare n RE-columns of degree $r = \max\{l, m\}$, and connect the s_i output of the j-th RE-column to the a_i input of the $(j + 1)$-st RE-column ($i \in \{1, \ldots, r\}, j \in \{1, \ldots, n - 1\}$). Also connect c_i to a_i of the first RE-column ($i \in \{1, \ldots, l\}$), and s_i of the n-th RE-column to d_i ($i \in \{1, \ldots, m\}$). For all q_h, q_j, c_i, and d_k, if $\delta(q_h, c_i) = (q_j, d_k)$, then connect the west output of the RE at (i, h) to the east input of the RE at (k, j). By this, M is simulated, and we obtain the following theorem.

Theorem 4.8 ([56]). *An RE is intrinsically universal.*

4.3.4.2 *Intrinsic universality of other 2-state RLEMs*

Here, we investigate intrinsic universality of 2-state RLEMs other than RE. First, consider the case where the number of input/output symbols is three or more. In [54] it has been shown that all non-degenerate 2-state k-symbol RLEM are intrinsically universal if $k \geq 3$. The following four lemmas have been proved to show the result.

Lemma 4.6 ([54, 57]). *RE is realizable by RLEM 3-10.*

By this lemma, we see RLEM 3-10 is intrinsically universal.

Lemma 4.7 ([57]). *RLEM 3-10 is realizable by RLEMs 2-3 and 2-4.*

Therefore, the set {RLEM 2-3, RLEM 2-4} is intrinsically universal.

Lemma 4.8 ([54]). *RLEMs 2-3 and 2-4 are realizable by any one of 14 non-degenerate 2-state 3-symbol RLEMs.*

Thus, any non-degenerate 3-symbol RLEM is intrinsically universal.

Lemma 4.9 ([54]). *Let M be an arbitrary non-degenerate 2-state k-symbol RLEM ($k \geq 3$). Then, there exists a non-degenerate 2-state $(k-1)$-symbol RLEM M' that is realizable by M.*

From Lemmas 4.6–4.9, we have the following theorem.

Theorem 4.9 ([54]). *All non-degenerate 2-state k-symbol RLEMs are intrinsically universal if $k \geq 3$.*

Lemmas 4.6–4.9 are proved by giving circuits that realize the target RLEMs.

First, consider Lemma 4.6. Figure 4.17 shows a circuit that realizes an RE [54]. This figure corresponds to the state H. By complementing the states of the bottom four RLEMs, we have the state V of an RE.

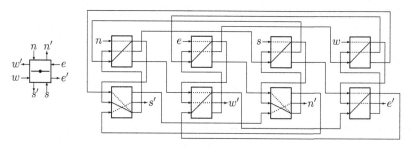

Fig. 4.17   A circuit composed of RLEM 3-10 that realizes RE [54].

It is easy to see that if a signal is given to the port e (w, respectively) in Fig. 4.17, then it goes out from the port w' (e') in two steps without changing the state of the circuit. This is a parallel case (Fig. 4.8(a)). On the other hand, if a signal is given to the port n, then the circuit evolves as shown in Fig. 4.19. The signal finally goes out from the port w', and the states of the bottom four RLEMs are complemented. This is an orthogonal case (Fig. 4.8(b)). In such a way, the circuit correctly simulates an RE.

Lemma 4.7 is proved by Fig. 4.18 [57]. It shows a circuit composed of RLEMs 2-3 and 2-4 that realizes RLEM 3-10.

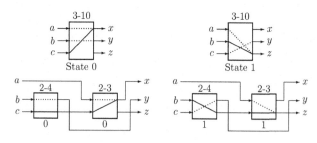

Fig. 4.18   RLEM 3-10 is realizable by RLEMs 2-3 and 2-4 [57].

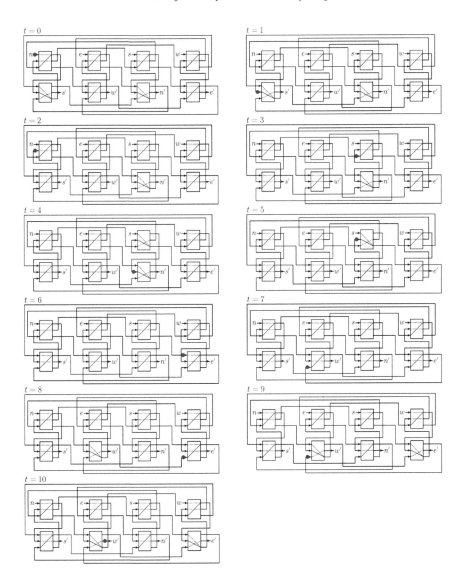

Fig. 4.19   Process of simulating RE by RLEM 3-10 when the state of RE changes.

Lemma 4.8 is proved by the 28 circuits shown in Fig. 4.20. The figure shows that RLEMs 2-2 and 2-3 are realizable by each of 14 non-degenerate 3-symbol RLEMs.

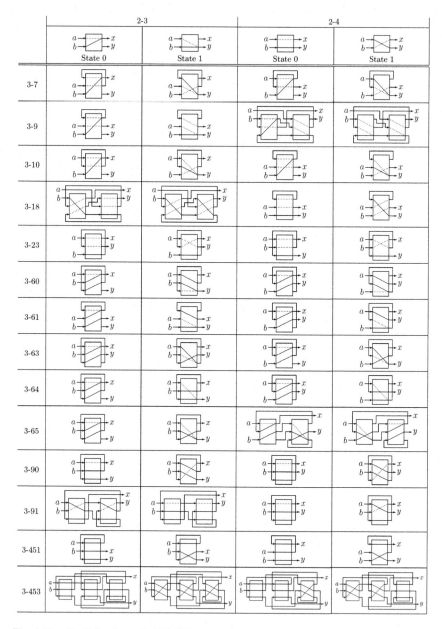

Fig. 4.20   Realizing 2-symbol RLEMs 2-3 and 2-4 by each of 14 kinds of non-degenerate 3-symbol RLEMs [54].

As for Lemma 4.9, we explain only a key idea of the proof. For a given k-symbol RLEM, we choose one output line and one input line, and connect them to make a feedback loop to obtain a $(k-1)$-symbol RLEM. Figure 4.21 shows the case of 4-symbol RLEM 4-23617. If we give an appropriate feedback loop, we can get a non-degenerate 3-symbol RLEM (upper row of Fig. 4.21). But, if we give an inappropriate feedback, then the resulting 3-symbol RLEM is a degenerate one (lower row of Fig. 4.21). In [54], it is proved that for a given non-degenerate k-symbol RLEM ($k > 2$), we can always find a feedback loop by which a non-degenerate $(k-1)$-symbol RLEM can be obtained.

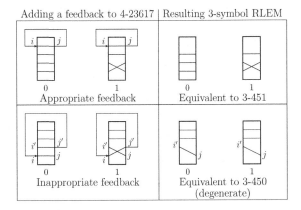

Fig. 4.21   Making a 3-symbol RLEM by adding a feedback loop to RLEM 4-23614.

Next, we investigate four non-degenerate 2-state 2-symbol RLEMs (see Fig. 4.14). They are RLEMs 2-2, 2-3, 2-4 and 2-17. The following four lemmas are shown in [9, 55] (their proofs are omitted here).

Lemma 4.10. *Neither RLEM 2-3, 2-4 nor 2-17 is realizable by RLEM 2-2.*

Lemma 4.11. *Neither RLEM 2-4 nor 2-17 is realizable by RLEM 2-3.*

Lemma 4.12. *Neither RLEM 2-3 nor 2-17 is realizable by RLEM 2-4.*

Lemma 4.13. *RLEM 2-2 is realizable by any one of RLEMs 2-3, 2-4 and 2-17.*

From Lemmas 4.10–4.13 we have the next theorem.

Theorem 4.10 ([55]). *RLEMs 2-2, 2-3 and 2-4 are not intrinsically universal. RLEM 2-2 is the weakest one among non-degenerate 2-state RLEMs.*

Recently, Matthew Cook and Ethan Palmiere proved that RLEM 2-17 is intrinsically universal. It is a remarkable result since the operation of RLEM 2-17 is quite simple.

Theorem 4.11 (Cook and Palmiere).[1] *RLEM 2-17 is intrinsically universal.*

Figure 4.22 summarizes the above intrinsic universality/non-universality results among non-degenerate 2-state RLEMs.

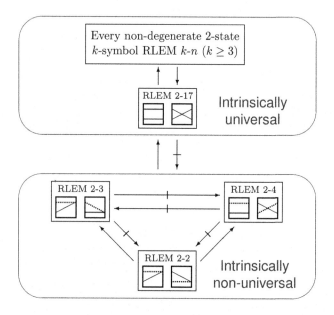

Fig. 4.22   Intrinsic universality/non-universality of non-degenerate 2-state RLEMs. Here, $A \to B$ ($A \not\to B$, respectively) indicates that A is (is not) realizable by B.

4.3.5 *Making RTMs out of RLEMs*

We construct 2-symbol RTMs with a rightward infinite tape using RLEMs. Though any intrinsically universal RLEM can be used to do so, we choose RE and RLEM 4-31. This is because RTMs are constructed easily by them.

---

[1]Personal communication.

4.3.5.1 *Composing RTMs using RE*

Since RE is intuitively easy to understand, we make RTMs out of it here. A composing method was first proposed in [51]. It was then revised in [9], and further revised in [27]. An RTM is constructed by assembling two kinds of functional modules. They are a tape cell module and a state module.

Tape cell module
A *tape cell module* simulates one tape square of a 2-symbol RTM. It is realized by a circuit shown in Fig. 4.23. Connecting infinite number of copies of it, a *tape unit* is obtained as shown in the right part of Fig. 4.28.

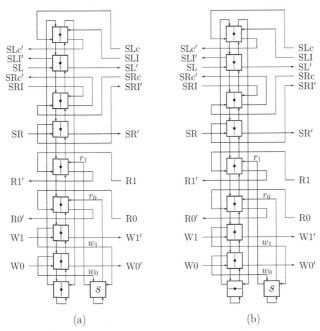

Fig. 4.23   Tape cell module composed of REs for 2-symbol RTMs. The state (a) shows that the head is not on this cell, and (b) shows that the head is on this cell.

The tape cell module keeps the information whether the head of the RTM is on this cell or not in its left part, which is an RE-column of degree 8 (Fig. 4.15). If the RE-column is in the state 0 (*i.e.*, its bottom RE is in the state V), then the head is not here (Fig. 4.23(a)). If it is in the state 1 (*i.e.*, its bottom RE is in the state H), then the head is here (Fig. 4.23(b)).

Reversible World of Cellular Automata

A tape symbol $s \in \{0, 1\}$ is stored in the RE indicated by s in Fig. 4.23, where 0 and 1 are represented by the states V and H, respectively. The right part of the tape cell is, actually, a *one-bit memory module* (Fig. 4.24) having the move function given in Table 4.6. It has two input ports w_0 and w_1, and two output ports r_0 and r_1. Assume the present state is $s \in \{0, 1\}$. If a signal is given to w_t ($t \in \{0, 1\}$), then the new symbol t is written in it, and the old symbol s is read-out from the output port r_s. Thus, a write operation always accompanies a read operation. It is a kind of a unit-time delay that keeps the last-received bit.

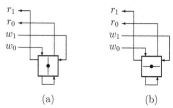

(a)   (b)

Fig. 4.24   One-bit memory module composed of RE. (a) State 0, and (b) state 1.

Table 4.6   Move function of the one-bit memory module in Fig. 4.24.

Present state	Input	
	w_0	w_1
0	0 r_0	1 r_0
1	0 r_1	1 r_1

Table 4.7   Ten kinds of symbols for the tape cell module and their meanings [9].

Symbol	Instruction/Response	Meaning
W0	Write 0	Instruction of writing the tape symbol 0 at the head position. By this instruction, read operation is also performed
W1	Write 1	Instruction of writing the tape symbol 1 at the head position. By this instruction, read operation is also performed
R0	Read 0	Response signal telling the read symbol at the head is 0
R1	Read 1	Response signal telling the read symbol at the head is 1
SL	Shift-left	Instruction of shift-left operation
SLI	Shift-left immediate	Instruction of placing the head on this cell by shifting left
SLc	Shift-left completed	Response (completion) signal of shift-left operation
SR	Shift-right	Instruction of shift-right operation
SRI	Shift-right immediate	Instruction of placing the head on this cell by shifting right
SRc	Shift-right completed	Response (completion) signal of shift-right operation

The tape cell module has ten input ports corresponding to ten kinds of input symbols listed in Table 4.7. They are interpreted as instructions to the tape unit or response signals to the finite control of an RTM. For each input symbol, there is a corresponding output symbol, which is indicated by the symbol with ', and thus the tape cell has ten input ports and ten output ports. A tape cell module can be defined as an RSM having ten input/output symbols (see Sect. 6.1.1 of [9]). However, here, we give intuitive explanations on its operations.

Making an infinite number of copies of a tape cell module, and connecting them to form a rightward infinite array, we can obtain a *tape unit* for the RTM. To the left of the tape unit a finite control of an RTM will be connected. We assume there is only one tape cell whose RE-column is in the state 1 in the initial setting, and thus there is only one head. Giving a signal to the tape unit, read/write and head-shift operations are performed.

First, consider the case where the head is not on this tape cell (Fig. 4.23(a)). Since its RE-column is in the state 0, a signal from the input port W0, W1, R0, R1, SL, SR, SLc, or SRc simply goes to the output port W0', W1', R0', R1', SL', SR', SLc', or SRc', respectively, without changing its state (see Table 4.4). It means that these signals skip tape cells having no head. Processing of a signal SLI or SRI in this case is discussed later.

Second, consider the case where the head is on this cell (Fig. 4.23(b)). The first subcase is that an input signal Wt ($t \in \{0,1\}$) is given, which is for writing the tape symbol t in this tape cell. We assume the one-bit memory module in its right part is in the state s. The signal changes the state of the RE-column to 0, and appears on the line w_t in Figs. 4.23 and 4.24. Then the state of the one-bit memory module changes to t, and the signal appears on the line r_s (see Table 4.6). This signal restores the RE-column to the state 1, and finally goes out from the port Rs'. Hence, the writing operation also performs a reading operation to keep reversibility of the tape cell. The signal Rs' moves leftward through tape cells having no head, and finally reaches the finite control of the RTM. Note that if an RTM needs to read a tape symbol, it is performed by sending a signal to the input port W0 of the tape unit. By this, the tape unit gives a response signal at the output port Rs', and the tape symbol at the head position is cleared to 0. Thus, this is a destructive readout.

The second subcase is that a signal is given to the input port SL of the tape cell with a tape head, which will shift the tape head to the left. This signal changes the RE-column to the state 0, and comes out from the port SLI'. The signal is then sent to the left-neighboring tape cell. If the latter tape cell receives an input signal SLI, then it sets the state of the RE-column to 1, and sends an output signal SLc' to the left. By such a process, shift-left operation is performed correctly.

The third subcase is that a signal is given to the input port SR of the tape cell with a tape head, which will shift the tape head to the right. The ports SR, SRI, and SRc are similar to the ports SL, SLI, and SLc, except that an output signal SRI′ is sent to the right-neighboring tape cell.

By above, we can see that read/write and head-shift operations are correctly performed by a tape unit.

State module

Before introducing a state module we first explain a subroutine call mechanism. Consider an RLEM circuit, which may be finite or infinite. A *subroutine* in the RLEM circuit is a black box having at least one calling (*i.e.*, input) port, and at least one return (*i.e.*, output) port (Fig. 4.25) that satisfies the following: If a calling signal is given to one of the calling ports, a return signal eventually comes out from one of the return ports. No signal should be given to a calling port before a return signal for the previous calling signal comes out.

c^1 →
c^0 →
r^1 ←
r^0 ←

Black box

Fig. 4.25   Subroutine. Here, it has two input ports c^0 and c^1 for calling it from a main routine, and two output ports r^0 and r^1 for returning to the main routine.

Here, to make a subroutine and its calling mechanism easily out of RLEMs, we restrict both the numbers of calling ports and return ports to be at most two. However, it is not an essential restriction (but to increase the numbers the circuit will become slightly complex). If there are two calling ports, say c^0 and c^1, we can give two kinds of information 0 and 1, which are regarded as input arguments, to the subroutine. Likewise, if there are two return ports r^0 and r^1, we can obtain two kinds of information 0 and 1, which are regarded as output values, from the subroutine.

A tape unit acts as three subroutines by suitably specifying calling and return ports. The first one is the subroutine having the calling ports W0 and W1, and the return ports R0′ and R1′. The second has the calling port SL, and the return port SLc′. The third has the calling port SR, and the return port SRc′.

A *subroutine caller* is a mechanism for calling one subroutine from many points of a main routine. Figure 4.26 is a subroutine caller composed of REs for a subroutine having calling ports c^0 and c^1, and return ports r^0 and r^1, It consists of n *caller units*. In this figure, each caller unit is composed of one RE. However, if we use other RLEM, two or more RLEMs may be needed to compose a caller unit. Here, c_i^s ($i \in \{1, \ldots, n\}, s \in \{0, 1\}$) is the i-th calling port for the main routine with the input s, and r_i^t ($i \in \{1, \ldots, n\}, t \in \{0, 1\}$) is the i-th return port for the main routine with the output t.

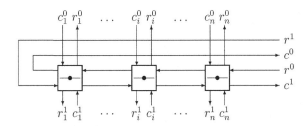

Fig. 4.26　Subroutine caller consisting of n caller units. It can be used from n points of a main routine.

Initially, all the REs in Fig. 4.26 are set to the state H. If a signal is given to the port c_i^s ($i \in \{1, \ldots, n\}, s \in \{0, 1\}$), then the state of the i-th RE changes to V, and the signal goes out from the port c^s. If the signal returns via the port r^t ($t \in \{0, 1\}$), then the state of the i-th RE is restored to H, and then the signal goes out from the port r_i^t. In this way, n points of the main routine can share the same subroutine. Note that if a subroutine has only one calling port or only one return port, then unnecessary lines in the caller are removed.

A *state module* simulates one state, say q_i, of an RTM. It is shown in Fig. 4.27. It is possible to formalize it as an RSM. But, here, we explain its operations based on its circuit diagram.

A state module consists of three submodules, which are *write-and-merge*, *head-shift*, and *read-and-branch* submodules. This figure shows the case where q_i is a right-shift state. The case for a left-shift state is similar. Because of the reversibility condition (Definition 4.3), shift direction is uniquely determined by the state. Note that a state module for an initial state consists only of a read-and-branch submodule, and that for a halting state consists of write-and-merge and head-shift submodules.

If the number of states of an RTM is m, then prepare m state modules, and connect them in a row to make a finite control of the RTM. At the

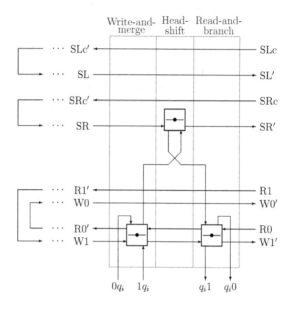

Fig. 4.27   State module composed of REs for a right-shift state q_i of an RTM.

left end of the array of the state modules, SLc′, SRc′, R1′ and R0′ are connected to SL, SR, W1 and W0, respectively.

Write-and-merge submodules and read-and-branch submodules in the m state modules form a subroutine caller that share the subroutine having the calling ports W0 and W1, and the return ports R0′ and R1′ in the tape unit. By this, read/write operations on the tape unit can be performed in each state. Head-shift submodules of the right-shift states also form a subroutine caller that share the subroutine having the calling port SR, and the return port SRc′ in the tape unit. Likewise, head-shift submodules of the left-shift states form a subroutine caller that share the subroutine having the calling port SL, and the return port SLc′.

Let $T = (Q, \{0, 1\}, q_0, F, 0, \delta)$ be a 2-symbol RTM. First consider how the write-and-merge submodule works. Assume $[p, s, 0, d, q_i], [p', s', 1, d, q_i]$ $\in \delta$. Note that it also works well for the case only one of these two quintuples exists. We further assume that the write-and-merge operation is done just after a read-and-branch operation that performs a destructive readout. Thus, the tape symbol at the head position is now 0. If the submodule receives a signal from the input port $0q_i$ ($1q_i$, respectively), then it sends a calling signal to the subroutine from the port W0′ (W1′) to perform a writing operation. Since the old symbol at the head position is 0, the submodule receives a signal from the return port R0 in both cases of writing 0 and 1. By this, two different signal paths of writing 0 and 1 are reversibly merged into one, and the signal is sent to the head-shift submodule.

The head-shift submodule works as follows. If it receives a signal from the write-and-merge submodule, it sends a signal to the calling port SL' or SR', by which shifting is performed in the tape unit. It receives a signal from the return port SLc or SRc. Then, the signal is sent to the read-and-branch submodule.

If the read-and-branch submodule receives a signal from the head-shift submodule, it sends a signal to the calling port W0' of the tape unit. Then the tape unit sends back a response signal via the return port R0 or R1 depending on the read symbol. The submodule finally gives a signal to the port q_i0 or q_i1. By above, read-and-branch operation is performed. Note that the read-and-branch submodule has a symmetric structure with that of the write-and-merge submodule.

State transitions of the RTM is realized by connecting state modules in the following way. If there is a quintuple $[q_i, s, t, d, q_j] \in \delta$, then the output port $q_i s$ of the state module for q_i is connected to the input port $t q_j$ of the state module for q_j.

Making RTMs

For a given m-state 2-symbol RTM M, prepare m state modules. Placing the modules in a row, and connecting them as described above, we have a finite control of M. Then, connect a tape unit to the right of the finite control. By this, we obtain a full circuit that simulates M. The resulting circuit is rightward-infinite. However, except a finite part, an exactly the same tape cell module, which has no head and contains the tape symbol 0, repeats indefinitely to the right. Thus, the circuit can be described finitely.

The circuit for the RTM T_{parity} in Example 4.1 is shown in Fig. 4.28. If we give a signal to the port "Start," then it is sent to the state module for the initial state. By this, T_{parity} begins to compute. If T_{parity} halts, then the signal from the state module corresponding to the accepting or rejecting state is sent to the port "Accept" or "Reject" showing that the computation is completed.

Figure 4.29 shows a screenshot of the simulator constructed on Golly, where the RE circuit of Fig. 4.28 is simulated [21]. There, circuits for RTMs T_{power} and T_{square} in Examples 4.2 and 4.3, and an RTM T_{prime}, which tests if a unary input is a prime, are also included.

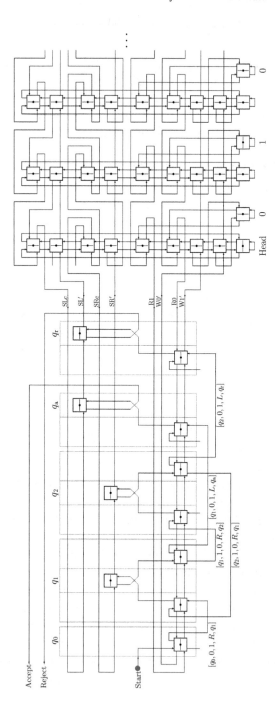

Fig. 4.28   RTM T_{parity} composed of REs.

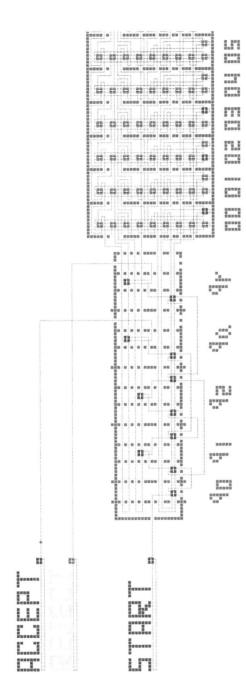

Fig. 4.29   Accepting configuration of RTM T_{parity} made of REs simulated on Golly [21].

4.3.5.2   *Composing RTMs using RLEM 4-31*

RLEM 4-31 is also useful for composing RTMs. One reason is that a tape cell module and a state module can be constructed using similar numbers of RLEMs as in the case of RE. More precisely, the numbers of elements for a tape cell module and a state module in the case of RLEM 4-31 are 9 and 5, while those in the case of RE are 10 and 3. There is also another reason. The move function of RLEM 4-31 is shown graphically in Fig. 4.30. For each of the states 0 and 1, there is only one case where the state changes to the other (in other words, each state has only one solid line). Generally, simulating a state-change of an RLEM in a reversible CA requires a complex mechanism. By this reason RLEM 4-31 can be implemented simply in some reversible CAs as we shall see in Chap. 7.

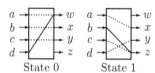

Fig. 4.30   RLEM 4-31.

Tape cell module

Figure 4.31 shows a *tape cell module* composed of RLEM 4-31. It realizes the same RSM as the one composed of RE (Fig. 4.23). Since it is designed in a similar manner as in the case of RE, we explain only the different points. Among the nine RLEMs in the tape cell module, the top one keeps the tape symbol, while the remaining eight keep the head position and process the ten kinds of instruction/response signals given in Table 4.7.

The top RLEM works as a *one-bit memory module* to memorize a tape symbol. If we denote its input ports by w_0 and w_1, and its output ports by r_0 and r_1 as in Fig. 4.32, it has the same move function given in Table 4.6.

Next, we explain the role of the lower eight RLEMs. If all the eight are in the state 0, then we regard that the head is not on this tape cell (Fig. 4.31(a)). If this is the case, a signal given to either W0, W1, R0, R1, SL, SLc, SR, or SRc goes through this module without changing its state, and comes out from the corresponding output port. When a signal is given to the port SRI (SLI, respectively), it makes all the eight go to the state 1, indicating that the head is now at this position, and finally comes out from the port SRc′ (SLc′).

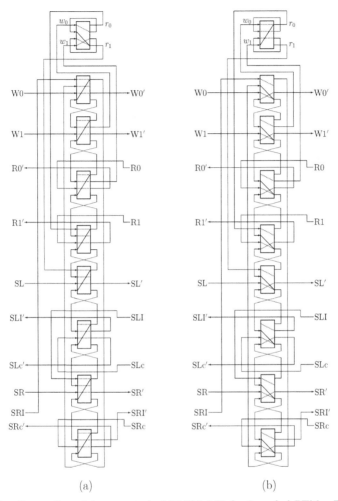

(a)   (b)

Fig. 4.31   Tape cell module composed of RLEM 4-31 for 2-symbol RTMs. The state (a) shows that the head is not on this cell and the tape symbol is 1, while (b) shows that the head is on this cell and the tape symbol is 0.

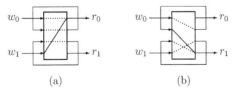

(a)   (b)

Fig. 4.32   One-bit memory module composed of RLEM 4-31. It is placed at the top of a tape cell module. (a) State 0, and (b) state 1.

If all the eight RLEMs are in the state 1, then we assume the head is on this tape cell (Fig. 4.31(b)). The first subcase is that a signal is given to the port Wt ($t \in \{0,1\}$). It first changes the states of the eight RLEMs to 0. Then, it goes to the input port w_t of the one-bit memory module. If the old state of the memory is $s \in \{0,1\}$, then the signal comes out from r_s. It restores the states of the eight RLEMs to 1, and finally comes out from the port Rs. By above, read/write operation is completed.

The second subcase is that a signal is given to the port SL. It changes the states of the eight RLEMs to 0. Then it comes out from SLI$'$. By this, the head shifts to the left-neighboring tape cell. The third subcase where a signal is given to SR is similar to the case of SL.

Making a rightward infinite array of tape cells, and connecting them, a tape unit is obtained as in the case of using RE.

State module
Figure 4.33 shows a *state module* composed of RLEM 4-31. A difference from the module composed of RE (Fig. 4.27) is that each of the write-and-merge and read-and-branch submodules has two RLEMs. These two play the role of one RE in the corresponding submodule in Fig. 4.27. Composition of a finite control of an RTM is similarly done.

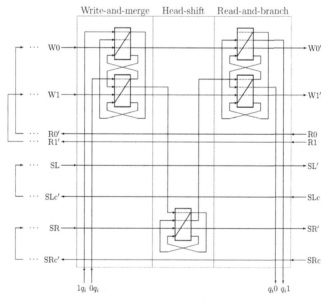

Fig. 4.33 State module composed of RLEM 4-31 for a right-shift state q_i of an RTM.

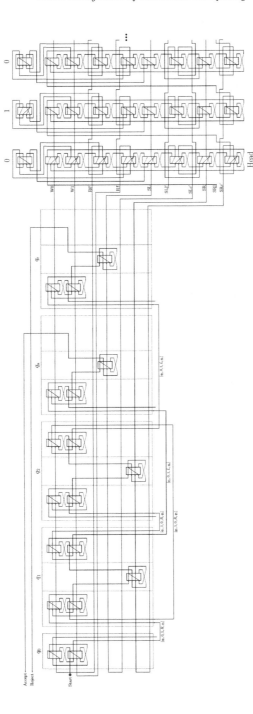

Fig. 4.34 RTM T_{parity} composed of RLEM 4-31 [21].

Making RTMs

Assembling the tape cell modules and state modules given above, we can compose any 2-symbol RTM. Figure 4.34 shows the circuit for the RTM T_{parity} in Example 4.1.

4.3.6 *Turing universality of RLEM*

Here, we examine why the notion of intrinsic universality of RLEMs was given as in Definition 4.19. We then introduce Turing universality, which is another notion of universality of RLEM.

Definition 4.19 states that if an RLEM R is intrinsically universal, then any RSM, in particular any RLEM M, can be realized by a circuit C composed only of R. Note that the input/output of the circuit C must be of the form of Fig. 4.6(b). It means that for any given circuit C_M composed of M, we can obtain a circuit C_R composed only of R that is equivalent to C_M, by replacing every occurrence of M in C_M by the circuit C.

Such a definition of intrinsic universality has some similarity to the theory of traditional combinatorial logic circuits. Note that, there, universality is often called *functional completeness*. It is well known that the sets of logic gates {AND, NOT} and {NAND} are complete sets for realizing all Boolean functions. For example, {NAND} is a complete set, since AND, OR and NOT are easily realized by circuits composed of NAND. Hence, for any given circuit composed of AND, OR and NOT we can make a circuit that realizes the same logical function by replacing AND, OR and NOT gates by the corresponding circuits made of NAND.

On the other hand, it is also known that {AND, OR} is not a complete set. However, if we employ a *dual-rail logic circuit* rather than a traditional *single-rail logic circuit*, we can realize *any* Boolean function using the set of single-rail logic gates {AND, OR}. In a dual-rail logic circuit, a Boolean value $A \in \{0, 1\}$ is represented by a pair of signals $(a, a') \in \{(0, 1), (1, 0)\}$, where $(a, a') = (0, 1)$ $((1, 0)$, respectively) represents $A = 0$ $(A = 1)$. Apparently, dual-rail NOT of A is obtained by exchanging a and a'. Dual-rail NAND is implemented by using single-rail AND and OR as shown in Fig. 4.35. Therefore, if we extend the notion of implementing Boolean functions, even an incomplete set of logic gates in the standard definition can be complete in a broader sense.

This is also the case in RLEMs. As we saw in Sec. 4.3.4, if an RLEM R is intrinsically universal, any RSM of the form of Fig. 4.6(b) is constructed out of R. Since a tape cell module and a state module can be

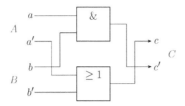

Fig. 4.35   Implementing dual-rail NAND $C = \overline{A \cdot B}$ using single-rail AND and OR, which are indicated by "&" and "≥ 1," respectively.

formalized as RSMs, RTMs are obtained as shown in Sec. 4.3.5. Intrinsic universality in Definition 4.19 is thus a "sufficient" condition for composing RTMs. However, it is not a "necessary" condition. There may be other methods of composing a circuit that simulates RTMs out of an RLEM that is not intrinsically universal. To include such possibilities, we define Turing universality of an RLEM as follows.

Definition 4.20. An RLEM R is called *Turing universal*, if the following condition holds: For any RTM T we can compose an infinite but finitely describable circuit C out of R that simulates T.

We should note that the circuit C must be infinite. Otherwise, it is a kind of a finite automaton. Here, we pose a restriction that C is *finitely describable*. The circuits of Figs. 4.28 and 4.34 are typical examples. They have a structure with infinite repetitions of a finite circuit except a finite part. This restriction comes from the assumption that we consider only finite computational configuration (ID) of an RTM T whose tape contains only blank symbols except a finite part (see Definition 4.4). However, here, we do not give a precise definition on the notion of finitely describable, since their formulation becomes cumbersome (note that in Sec. 4.4.1 a similar notion on infinite configurations of CAs is given).

Matthew Cook and Ethan Palmiere proved an *amazing* result showing that any RTM can be simulated by an infinite but finitely describable circuit composed only of RLEM 2-2, which is not intrinsically universal. Their method of composing RTMs out of RLEM 2-2 is quite complex, but its correctness is verified by computer simulation.

Theorem 4.12 (Cook and Palmiere). [2] *RLEM 2-2 is Turing universal.*

---

[2] Personal communication.

Since RLEM 2-2 is the weakest 2-state RLEM (see Fig. 4.22), all 2-state non-degenerate RLEMs are Turing universal. Therefore, Turing universality is a weaker notion than intrinsic universality (but, actually, it is not so weak since any RTM can be constructed out of a Turing universal RLEM). Note that RLEM 2-2 is equivalent to the one-bit memory module shown in Table 4.6.

4.3.7   *Advantage of using RLEM*

In the conventional design theory and practice of logic circuits, logic gates such as AND, OR, NOT, NAND are used as primitives. Thus, logic gates are dealt with separately from memory elements. On the other hand, as we have seen above, in the theory of reversible computing RLEMs are useful, since RSMs and RTMs are constructed rather simply out of RLEMs.

Another advantage of using RLEMs is that there is no need to synchronize incoming signals at each element. In the case of a logic gate with two or more input ports, input signals must arrive at the gate exactly the same time. It is, of course, possible to construct an RTM using only reversible logic gates. However, if we do so, adjustment of the signal timing at each gate becomes very cumbersome, since simple waiting of other signal is not possible in a reversible system. On the other hand, in an RLEM, an incoming signal interacts with the state of the RLEM. Therefore, the input signal can be given at any moment, and the operation of the RLEM is triggered by it. This property greatly simplifies a construction of RTMs out of RLEMs.

4.4   Turing Universality of PCA

In the following chapters, we study various simple reversible PCAs in which reversible computers can be embedded. We now discuss how Turing universality is defined in PCAs. Basically, a PCA is called Turing universal if it can simulate any Turing machine. Here, we allow infinite configurations to simulate them. In this case, however, configurations must be finitely describable. Otherwise, we cannot specify an initial configuration of a PCA effectively. We first give a suitable class of finitely describable infinite configurations, and then define Turing universality of a PCA having such infinite configurations. We also give a definition of Turing universality for a PCA having only finite configurations, which is a stronger version of Turing universality.

4.4.1 *Horizontally ultimately periodic configuration*

We define a special type of configurations for SPCAs. Since a similar definition is possible for TPCAs, its definition is omitted here.

Definition 4.21. Let $P = (\mathbb{Z}^2, (T, R, B, L), ((0, -1), (-1, 0), (0, 1), (1, 0)),$ $f, \#)$ be an SPCA, where $\# \in Q = T \times R \times B \times L$ is a quiescent state of P. A configuration $\alpha : \mathbb{Z}^2 \to Q$ is said to be *horizontally ultimately periodic* if there are $p_L, p_R \in \mathbb{N} - \{0\}$, $x_L, x_R \in \mathbb{Z}$ $(x_L < x_R)$, and $y_0, y_1 \in \mathbb{Z}$ $(y_0 \leq y_1)$ that satisfy the following.

$$\forall (x, y) \in \mathbb{Z}^2 : (((y_0 \leq y \leq y_1) \wedge (x \leq x_L) \Rightarrow \alpha(x - p_L, y) = \alpha(x, y)) \wedge$$
$$((y_0 \leq y \leq y_1) \wedge (x \geq x_R) \Rightarrow \alpha(x, y) = \alpha(x + p_R, y)) \wedge$$
$$((y < y_0 \vee y > y_1) \Rightarrow \alpha(x, y) = \#))$$

The positive integers p_L and p_R are called the *left period* and the *right period* of the configuration, respectively. The set of all horizontally ultimately periodic configurations of P is denoted by $\mathrm{Conf}_{\mathrm{hup}}(P)$.

By the above definition, we can see that a horizontally ultimately periodic configuration is one that satisfies the following (see Fig. 4.36).

(1) There are three finite arrays of states

$$S_L = \begin{matrix} a_{1,1} \cdots a_{1,p_L} \\ \vdots \qquad \vdots \\ a_{h,1} \cdots a_{h,p_L} \end{matrix}, \quad S_C = \begin{matrix} b_{1,1} \cdots b_{1,n} \\ \vdots \qquad \vdots \\ b_{h,1} \cdots b_{h,n} \end{matrix}, \text{ and } \quad S_R = \begin{matrix} c_{1,1} \cdots c_{1,p_R} \\ \vdots \qquad \vdots \\ c_{h,1} \cdots c_{h,p_R} \end{matrix}$$

where $h = y_1 - y_0 + 1$ and $n = x_R - x_L - 1$.

(2) The array S_L repeats infinitely to the left, the array S_R repeats infinitely to the right, and the array S_C lies between them.
(3) All the other cells are in the quiescent state $\#$.

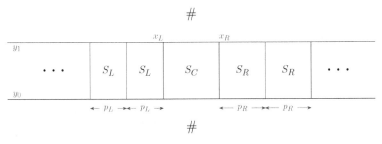

Fig. 4.36   Horizontally ultimately periodic configuration.

A *finite configuration* is a horizontally ultimately periodic configuration that satisfies $p_L = p_R = 1$ and $a_{i,1} = c_{i,1} = \#$ for all $i \in \{1, \ldots, h\}$, *i.e.*, S_L and S_R are blank arrays. We write the set of all finite configurations of P of a PCA (*i.e.*, SPCA or TPCA) by $\mathrm{Conf}_{\mathrm{fin}}(P)$.

Let P be a PCA, and let $\alpha \in \mathrm{Conf}_{\mathrm{hup}}(P)$. Then, as shown above, α is *finitely describable* by the three arrays S_L, S_C, and S_R. We call the following $D(\alpha)$ a *description* of α.

$$D(\alpha) = (S_L, S_C, S_R)$$

The set of all descriptions of configurations in $\mathrm{Conf}_{\mathrm{hup}}(P)$ is denoted by $D(\mathrm{Conf}_{\mathrm{hup}}(P))$. Likewise, the set of all descriptions in $\mathrm{Conf}_{\mathrm{fin}}(P)$ is denoted by $D(\mathrm{Conf}_{\mathrm{fin}}(P))$.

$$D(\mathrm{Conf}_{\mathrm{hup}}(P)) = \{D(\alpha) \mid \alpha \in \mathrm{Conf}_{\mathrm{hup}}(P)\}$$
$$D(\mathrm{Conf}_{\mathrm{fin}}(P)) \; = \{D(\alpha) \mid \alpha \in \mathrm{Conf}_{\mathrm{fin}}(P)\}$$

Besides the configurations in $\mathrm{Conf}_{\mathrm{hup}}(P)$, there are many other finitely describable infinite configurations. In this book, however, we consider only those in $\mathrm{Conf}_{\mathrm{hup}}(P)$. This is because it is sufficient to use configurations in $\mathrm{Conf}_{\mathrm{hup}}(P)$ for simulating RLEM circuits that realize RTMs (as in Figs. 4.28 and 4.34) as we shall see in the following chapters.

4.4.2   *Defining Turing universality of PCA*

We first define the notion of simulation where a computing system is simulated in a PCA. Note that, here, the PCA is not necessarily reversible.

Let M be a computing system (*e.g.*, TM, CM, PCA, *etc.*). We denote the set of all instantaneous descriptions of M by $\mathrm{ID}(M)$. If M is a PCA, then we use $D(\mathrm{Conf}_{\mathrm{hup}}(M))$ (or $D(\mathrm{Conf}_{\mathrm{fin}}(M))$ when simulating finite configurations of M) instead of $\mathrm{ID}(M)$.

Definition 4.22. Let M be a computing system, and P be a PCA whose global function is F_P. Let $\alpha \in \mathrm{ID}(M)$. If there exist a monotonically increasing total recursive function $\tau_\alpha : \mathbb{N} \to \mathbb{N}$, and two total recursive functions $c : \mathrm{ID}(M) \to D(\mathrm{Conf}_{\mathrm{hup}}(P))$, and $d : D(\mathrm{Conf}_{\mathrm{hup}}(P)) \to (\mathrm{ID}(M) \cup \{\bot\})$ that satisfy the following, then we say P *with horizontally ultimately periodic configurations simulates* M. Note that $d(\beta) = \bot$ for $\beta \in D(\mathrm{Conf}_{\mathrm{hup}}(P))$ means there is no $\alpha \in \mathrm{ID}(M)$ that corresponds to β.

$$\forall \alpha, \alpha' \in \mathrm{ID}(M), \; \forall n \in \mathbb{N} \; (\alpha \vdash_{M}^{n} \alpha' \;\Rightarrow\; d(F_A^{\tau_\alpha(n)}(c(\alpha))) = \alpha')$$

Here, c and d are *encoding* and *decoding functions*, respectively. As a special case, if c and d are total recursive functions such that $c : \mathrm{ID}(M) \to D(\mathrm{Conf}_{\mathrm{fin}}(P))$, and $d : D(\mathrm{Conf}_{\mathrm{fin}}(P)) \to (\mathrm{ID}(M) \cup \{\perp\})$, then P *with finite configurations* simulates M. Note that, generally, the function τ_α depends on the initial ID α of M. However, if $\tau_\alpha = \tau_\beta$ holds for all $\alpha, \beta \in \mathrm{ID}(M)$, then we write them by τ.

Definition 4.23. We say that a PCA P (which is not necessarily reversible) with horizontally ultimately periodic configurations (finite configurations, respectively) is *Turing universal* if P with horizontally ultimately periodic configuration (finite configuration) can simulate *any* TM.

Turing universality of P with finite configuration is a stronger notion than the other. However, hereafter, the description "with horizontally ultimately periodic configurations" will be omitted in the case of weaker universality, since only the SPCA P_3 in Sec. 8.5 is strongly universal.

4.5 Intrinsic Universality of PCA

Intrinsic universality of a CA U is a property where *any* CA P in some large class $\mathcal{P}$ of CAs is simulated by U (see, *e.g.*, [36,58]). It should be noted that U must be able to simulate any evolution process of configurations even if they are infinite configurations. We first make some preparations to define intrinsic universality of an SPCA (it is similarly defined for a TPCA).

Let P_1 and P_2 be SPCAs:

$$P_1 = (\mathbb{Z}^2, (T_1, R_1, B_1, L_1), ((0, -1), (-1, 0), (0, 1), (1, 0)), f_1)$$
$$P_2 = (\mathbb{Z}^2, (T_2, R_2, B_2, L_2), ((0, -1), (-1, 0), (0, 1), (1, 0)), f_2)$$

Let $Q_i = T_i \times R_i \times B_i \times L_i$ ($i \in \{1, 2\}$), $h, w \in \mathbb{N} - \{0\}$, and $A = \{0, \ldots, w - 1\} \times \{0, \ldots, h - 1\}$. Let $c : Q_1 \to Q_2^A$ be an injection (hence h and w must be the ones such that $|Q_1| \le |Q_2^A|$), and $d : Q_2^A \to (Q_1 \cup \{\perp\})$ be the function induced by c defined below.

$$\forall X \in Q_2^A : \left(d(X) = \begin{cases} q & \text{if } c(q) = X \\ \perp & \text{elsewhere} \end{cases} \right)$$

We call c an *encoding function* that maps a state $q \in Q_1$ to a rectangular array $X \in Q_2^A$. On the other hand, d is a *decoding function* that maps a rectangular array $X \in Q_2^A$ to the state $q \in Q_1$ such that $c(q) = X$. Note that $d(X) = \perp$ if there is no q such that $c(q) = X$.

The functions c and d are extended to $c^* : \mathrm{Conf}(P_1) \to \mathrm{Conf}(P_2)$ and $d^* : \mathrm{Conf}(P_2) \to (\mathrm{Conf}(P_1) \cup \{\bot\})$ as follows.

$\forall \alpha_1 \in \mathrm{Conf}(P_1), \forall \alpha_2 \in \mathrm{Conf}(P_2) :$

$\quad c^*(\alpha_1) = \alpha_2$ if $\forall(x_1, y_1) \in \mathbb{Z}^2,\ \forall(x_2, y_2) \in A :$

$$(\alpha_2(wx_1 + x_2, hy_1 + y_2) = c(\alpha_1(x_1, y_1))(x_2, y_2))$$

$\forall \alpha_2 \in \mathrm{Conf}(P_2), \forall \alpha_1 \in \mathrm{Conf}(P_1) :$

$$d^*(\alpha_2) = \begin{cases} \alpha_1 \text{ if } \forall(x_1, y_1) \in \mathbb{Z}^2,\ \forall(x_2, y_2) \in A : \\ \qquad (\alpha_2(wx_1 + x_2, hy_1 + y_2) = c(\alpha_1(x_1, y_1))(x_2, y_2)) \\ \bot \text{ elsewhere} \end{cases}$$

By above, we can see that c^* and d^* are finitely defined using c and d. Hereafter, c^* and d^* are also called an *encoding function*, and a *decoding function*, respectively.

Figure 4.37 shows how a configuration α_1 of P_1 is encoded and embedded in $\alpha_2 = c^*(\alpha_1)$ of P_2 by the function c^*.

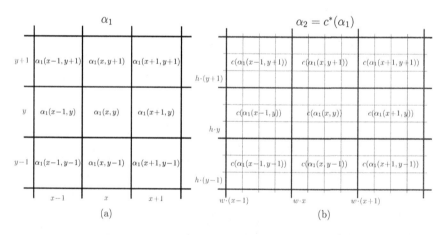

Fig. 4.37   Embedding a configuration α_1 of the SPCA P_1 into a configuration α_2 of the SPCA P_2 by the encoding function c^*. (a) The configuration α_1. (b) The configuration α_2, in which each cell's state of P_1 is represented by a state of a metacell, which is the array of states of P_2's cells of size $h \times w$ obtained by the function c.

Now, intrinsic universality of SPCA is defined as follows.

Definition 4.24. Let $\mathcal{P}$ be a class of SPCAs. An SPCA U, whose set of states and global function are Q_U and F_U, is called *intrinsically universal* with respect to $\mathcal{P}$, if the following condition holds.

$\forall P \in \mathcal{P}$, whose set of states and global function are Q_P and F_P,

$\exists\, h, w \in \mathbb{N} - \{0\},\ \exists \tau \in \mathbb{N} - \{0\}$,

$\exists\, c : Q_P \to Q_U^A$, an encoding function,

$\exists\, d : Q_U^A \to (Q_P \cup \{\bot\})$, a decoding function,

where $A = \{0, \ldots, w-1\} \times \{0, \ldots, h-1\}$,

$\forall \alpha, \alpha' \in \mathrm{Conf}(P) :$

$$(\ F_P(\alpha) = \alpha' \ \Rightarrow \ d^*(F_U^\tau(c^*(\alpha))) = \alpha' \ )$$

An array of U's cells on A is called a *metacell* for P. Thus, Q_U^A is the set of states of a metacell. Note that h, w, τ, c and d vary depending on P. If the above holds for an SPCA P, then U simulates one step of P in τ steps. Also note that if $\mathcal{P}$ contains a class of Turing universal SPCAs, then U is also Turing universal, as well as intrinsically universal.

In Sec. 6.4.4, it will be shown that the reversible ESPCA-02c5bf is intrinsically universal with respect to the class of all reversible SPCAs. There, an RE, a universal RLEM, is firstly implemented in ESPCA-02c5bf. Then, a metacell for a given reversible SPCA is constructed using REs.

Generally, if a universal RLEM (or a universal reversible logic gate) is realized in a reversible PCA, then it is intrinsically universal, since a cell of any other reversible PCA is realized using the logic element in a similar manner as in the case of ESPCA-02c5bf.

4.6 Remarks and Notes

In this chapter, three models of universal reversible computing systems are described. They are a reversible Turing machine (RTM), a reversible counter machine (RCM), and a reversible logic element with memory (RLEM). In the following chapters, they are used to show Turing universality of simple reversible PCAs.

An RTM is a standard model in the theory of reversible computing. Since the class of 2-symbol RTMs with a rightward infinite tape is Turing universal (Theorem 4.2), these RTMs will be used in the following.

An RLEM is a useful device to compose RTMs in reversible PCAs. If we use a logic gate as a logical device, two or more input signals must arrive at each gate exactly at the same time. On the other hand, in an RLEM, an incoming signal interacts with the states of the RLEM, and thus there is no need to synchronize signals. By this, construction of large circuits is greatly simplified. Furthermore, in Sec. 4.3.5, it is shown that any RTM can be constructed out of RE or RLEM 4-31 rather easily, which are typical universal RLEMs. Therefore, in the following chapters, these universal RLEMs are first realized in several simple reversible PCAs, and then RTMs are constructed out of them in their cellular spaces.

An RCM is another model of reversible computing. It has a finite number of counters as an auxiliary memory, each of which can keep a non-negative integer. It is known that the class of RCMs with only two counters is Turing universal [49] (Theorem 4.5). An RCM is useful to obtain a Turing universal reversible PCA having finite configurations (see Definition 4.23). The reason is that a counter can be implemented by a small pattern called a position marker in the PCA, where a non-negative integer is kept by the distance between the marker and the origin. Increment/decrement operation to a counter is performed by shifting a marker by a signal. The finite control of the RCM is constructed out of RLEMs as in the case of an RTM. Such a construction method will be shown in Sec. 8.4.

Besides Turing universality, intrinsic universality of a PCA is defined and discussed in Sec. 4.5. It is the property of a PCA where *any* CA in some large class of CAs can be simulated in the PCA. In Sec. 6.4.4, we shall see that reversible ESPCA-02c5bf is intrinsically universal with respect to all reversible PCAs. From the method shown there, we can also see that any reversible PCA is intrinsically universal if a universal RLEM, such as an RE, can be realized in it.

In [21], simulators for 2-symbol TMs, CMs, and RLEM circuits (as well as for PCAs) are given, which are executable on Golly [1]. Using these simulators, we can visually understand how these models work (as shown in Fig. 4.29). Although these computing models themselves are not cellular automata, it is very useful to have their simulators on Golly. This is because any TM, any CM, and any RLEM circuit can be simulated on them by giving appropriate initial configurations for them. In addition, even when their computing processes need huge numbers of steps and very large configurations, they can be simulated on Golly in a reasonable time.

4.7 Exercises

4.7.1 *Paper-and-pencil exercises*

4.7.1.1 *Exercise on RTMs*

Exercise 4.1.** Design the following RTMs, and verify that these RTMs are actually reversible.

(1) Compose an RTM T_{add} that performs addition of unary numbers. It does the following computation for any non-negative integers m and n, where q_0 and q_f are the initial and the final states of T_{add}, respectively.

$$q_0 01^m 01^n \mid^*_{\overline{T_{\mathrm{add}}}} q_f 01^m 01^{m+n}$$

(2) Compose an RTM T_{mult} that performs multiplication of unary numbers. It does the following computation for any non-negative integers m and n, where p_0 and p_f are the initial and the final states of T_{mult}, respectively.

$$p_0 01^m 01^n \mid^*_{\overline{T_{\mathrm{mult}}}} p_f 01^m 01^n 01^{mn}$$

Hint: Use T_{add} as a "subroutine."

4.7.1.2 *Exercise on RCMs*

Exercise 4.2.* Design the following RCMs.

(1) Compose an RCM(2) M_{half} in the quadruple form that performs the following computation for any given non-negative integer n, where h_0 and h_f are the initial and the final states, respectively. Also verify that M_{half} is actually a reversible CM(2).

$$(h_0, (n, 0)) \mid^*_{\overline{M_{\mathrm{half}}}} (h_f, (n \bmod 2, \ \lfloor n/2 \rfloor))$$

(2) Convert M_{half} into an RCM(2) M'_{half} in the program form with a WFP P_{half}. It performs the following computation for any non-negative integers n, where a_f is a final address of M'_{half}.

$$(0, (n, 0)) \mid^*_{\overline{M'_{\mathrm{half}}}} (a_f, (n \bmod 2, \ \lfloor n/2 \rfloor))$$

4.7.1.3 *Exercises on RLEMs*

Exercise 4.3.* Compose RLEM 3-7 (Fig. 4.38) out of REs by the systematic method shown in Sec. 4.3.4.1. Namely, design a circuit of REs that simulates RLEM 3-7 like the one in Fig. 4.16.

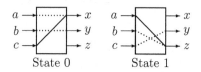

State 0   State 1

Fig. 4.38   RLEM 3-7.

Exercise 4.4.** Show that RLEM 3-7 can be realized by using only one RE. Hint: Add an appropriate feedback loop to an RE.

Exercise 4.5.*** Show that RLEM 4-31 can be realized by a circuit having only two copies of RLEM 3-7 (its answer is found in [59]). Therefore, from Exercise 4.4, RLEM 4-31 is realized by two REs.

Exercise 4.6.*** Show that RE can be realized by a circuit containing four copies of RLEM 3-7 (its answer is found in [59]).

Exercise 4.7.** Sketch a circuit composed of REs that simulates the RTM T_{add} in Exercise 4.1.

4.7.2 *Golly exercises*

Although RTMs, RCMs and RLEMs are not CAs, Golly is also useful for simulating them.

Exercise 4.8.** Simulate RTMs T_{add} and T_{mult} in Exercise 4.1 on Golly using the TM simulator contained in [21]. Observe how the RTMs work on the simulator.

Exercise 4.9.** Simulate RCMs M_{half} and M'_{half} in Exercise 4.2 on Golly using the CM simulators for the quadruple form and the program form contained in [21].

Exercise 4.10.** Simulate the circuits composed of RLEMs in Exercises 4.3–4.7 on Golly using the simulator contained in [21].
Note: Since RLEM 3-7 is not directly simulated in the simulator in [21], use the result of Exercise 4.4.

PART 2

Fantastic Phenomena and Computing in Reversible Cellular Automata

Chapter 5

Fantastic Phenomena in Reversible PCAs

Reversible ESPCAs and ETPCAs often exhibit amazing evolution processes despite their simplicity of local functions. Here, we investigate their behavior, and show various fascinating phenomena. As discussed in Chap. 2, there are three kinds of patterns in reversible PCAs. They are periodic, space-moving, and diameter-growing patterns. We shall find many interesting patterns in reversible ESPCAs and ETPCAs. Furthermore, interactions of these patterns often produce unexpected reactions. In this chapter, we first consider the Game of Life cellular automaton (GoL). Though GoL is an irreversible CA, it is worth looking at it, since it gives a clue to clarifying similarities and differences between irreversible CAs and reversible PCAs. We then investigate reversible PCAs, and show remarkable periodic and space-moving patterns, and their interactions. Patterns that continuously generate space-moving patterns (called a *gun*) can be found in various reversible ESPCAs and ETPCAs. Furthermore, many population-growing patterns show fantastic pattern generation capability. Because of T-symmetry of these reversible PCAs, we can find interesting inverse evolution processes of them. We also discuss linear ESPCAs and ETPCAs, which include reversible and irreversible ones, and show that fractal-like patterns and self-replicating ones are generated by them. Since there are many ESPCAs and ETPCAs, we discuss only selected ones which are particularly interesting. For those PCAs that have not been fully investigated, readers may find a variety of new fascinating phenomena.

5.1 Evolution in the Game of Life (GoL)

As it is explained in Example 1.2, the *Game of Life* (GoL) is a 2-dimensional cellular automaton with the Moore neighborhood, which was proposed by

J.H. Conway, and introduced by M. Gardner [4, 5]. Each cell of GoL has two states: a live state (represented by ●) and a dead state (represented by a blank). As noted in Example 1.2, its local function can be shortly described by the following: Just 3 for BIRTH, 2 or 3 for SURVIVAL [6].

So far, an enormous number of patterns of GoL that show interesting behavior have been shown. Here, we pick up only several of them that are related to the following sections. For other topics on GoL and their details, see LifeWiki [7], and the book by Johnston and Greene [8].

5.1.1 *Various patterns in GoL*

Figure 5.1 shows an example of a typical evolution process in GoL.

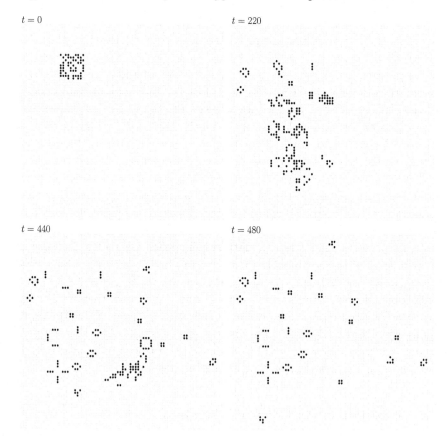

Fig. 5.1 Example of a typical evolution process of configurations in GoL starting from a small random pattern.

It starts from a random pattern of size 10×10 ($t = 0$). First, it becomes a configuration that contains actively changing regions ($t = 220$). Then, stable, periodic, and space-moving patterns are created ($t = 440$). Finally, actively changing regions disappear and only stable, periodic, and space-moving patterns remain ($t = 480$). At $t = 480$ we can find three space-moving patterns called *gliders*, which move to the north-west, south-west, and south-east directions. These configurations are diameter-growing ones as a whole, though the total number of live cells is kept constant after $t = 480$.

5.1.1.1 *Still life in GoL*

In GoL, special terminology is used for denoting periodic, and space-moving patterns. A stable pattern (*i.e.*, a periodic pattern of period 1) in GoL is called a *still life*. Figure 5.2 shows examples of still lifes that appear in Fig. 5.1. Of course, there are still more still lifes. Their details are described in [7, 8].

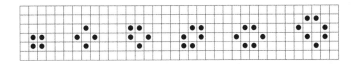

Fig. 5.2   Still lifes in GoL that appear in Fig. 5.1. They are called *block, tub, boat, ship, beehive*, and *loaf* (from left to right) [4].

5.1.1.2 *Oscillator in GoL*

A periodic pattern having a longer period than 1 is called an *oscillator*. In Fig. 5.1, only one kind of oscillator called a blinker (Fig. 1.6) appears. Figure 5.3 is another kind of oscillator called a *clock*. Many kinds of oscillators exist in GoL (see [7, 8]).

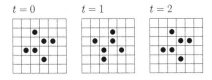

Fig. 5.3   *Clock* [4], an oscillator of period 2, in GoL.

5.1.1.3   *Glider and spaceship in GoL*

The pattern shown in Fig. 5.4 is called a *glider*. The glider moves by one cell diagonally in four steps. It is the most famous space-moving pattern in GoL. It was discovered by Richard K. Guy, and introduced in [4]. It not only flies in the cellular space, but also shows fascinating phenomena caused by the interactions with other patterns or other gliders. Using these phenomena, complex functional objects, such as universal computers, have been constructed.

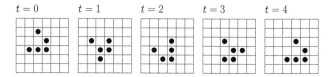

Fig. 5.4   *Glider* [4] in GoL.

Space-moving patterns other than the glider are called *spaceships*. For example, Fig. 5.5 is a pattern called a *lightweight spaceship* [6]. As for other spaceships see [7, 8].

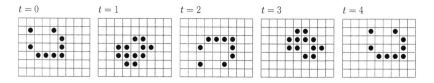

Fig. 5.5   Lightweight spaceship [6] in GoL.

Among many kinds of spaceships, only the glider has a particular name, since it has many interesting properties besides the fact it was discovered first. In this book, we also follow this convention: In each ESPCA or ETPCA, the most useful and/or remarkable space-moving pattern is called a glider, and the remaining ones are simply called space-moving patterns.

5.1.1.4   *Growing pattern in GoL*

In [4], the following conjecture by J.H. Conway was reported: No finite pattern can grow indefinitely in its population. Soon after that, a counter-example to this conjecture was given [5]. It is the *glider gun* created by Bill Gosper (Fig. 5.6). It generates a glider every 30 time steps. Thus, it is not only diameter-growing, but also population-growing.

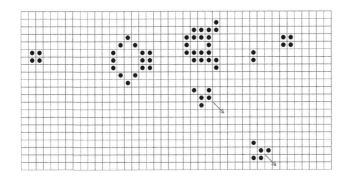

Fig. 5.6   The *glider gun* by B. Gosper [5].

Discovery of the glider and the glider gun was a particularly remarkable fact, and it made GoL very attractive. By this, many Life-enthusiasts appeared, and a huge number of fascinating phenomena in GoL have been found or created by them (see [7, 8]).

5.1.2 *Interactions of patterns in GoL*

Interactions of patterns of GoL often produce useful phenomena. Several examples are shown below.

5.1.2.1 *Interaction of gliders*

Figures 5.7 and 5.8 are examples of interactions of two gliders. In [6], it is shown that combining the erasing process given in Fig. 5.7 and the glider gun in Fig. 5.6, logical gates NOT, AND, and OR are realized.

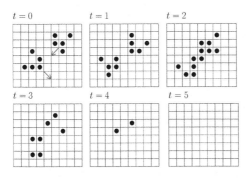

Fig. 5.7   Colliding two gliders in this way, both of them are erased [6].

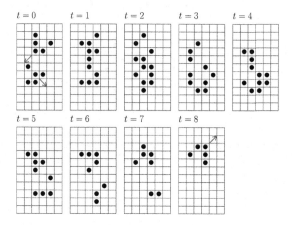

Fig. 5.8   Colliding two gliders in this way, one is erased and the other goes back [6]. This process is called the *kickback reaction*.

Note that in a reversible PCA, erasure of gliders such as in Figs. 5.7 and 5.8 is not possible, though erasure of one glider-6 among three shown in Fig. 5.69 is possible in ETPCA-0347.

5.1.2.2   *Interaction of gliders and a block*

Colliding two gliders with a block as in Fig. 5.9, the position of the block is pulled diagonally by three cells, and the gliders are dissolved. It is also possible to push a block by gliders [6]. Such phenomena can be used as a memory, which is called a *sliding block memory* where the memory states are kept by the position of the block. Namely, a block is used as a *position marker*. In the following, we shall use small periodic patterns as position markers to implement a memory in some reversible PCAs.

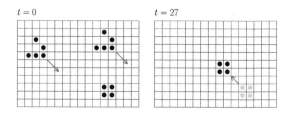

Fig. 5.9   Pulling a block by two gliders [6].

5.1.2.3 *Reflecting a glider*

In GoL, there are various patterns for reflecting a glider (see "Reflector" in [7]). Placing reflectors suitably, the move direction and the timing of a glider can be freely adjusted. Figure 5.10 is one of the stable reflectors called a *snark* proposed by M. Playle (see "Snark" in [7,8]). In some reversible PCAs, small reflectors for space-moving patterns exist. By them, we can use space-moving one as a *signal* for composing reversible Turing machines.

It is also noted that a *snark loop* [7] can be composed out of four rotated and unrotated copies of a snark, on which gliders go around. Thus, it is an oscillator as a whole. By adjusting the positions of four snarks, we have an oscillator of arbitrarily long period in GoL.

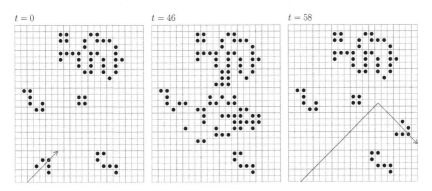

Fig. 5.10   90°-right-turn of a glider is possible by the pattern *snark* [7,8].

5.2   Periodic Patterns in Reversible PCAs

Now, we investigate reversible PCAs. We first give periodic patterns in several selected reversible PCAs, and show their properties.

5.2.1   *Periodic patterns in ESPCAs*

We consider two categories of reversible ESPCAs below. They are conservative ESPCAs and non-conservative ones.

5.2.1.1   *The case of reversible and conservative ESPCAs*

ESPCA-01wxyf   Consider the reversible and conservative ESPCA-01wxyf where $w \in \{3, 6, 9, c\}$, $x \in \{5, a\}$, and $y \in \{7, b, d, e\}$. By instantiating the variables w, x and y, we have 32 ESPCAs. Thus the expression

ESPCA-01wxyf stands for the 32 ETPCAs collectively. Figure 5.11 shows
the three local transition rules that are common to all the 32 ESPCAs.

Fig. 5.11   Partial local function of ESPCA-01wxyf.

First, we give periodic patterns having short periods. A *rotor*, a periodic
pattern of period 4 in ESPCA-01wxyf, is shown in Fig. 5.12. It evolves
exactly in the same way in all the 32 ESPCAs expressed by ESPCA-01wxyf.
Figure 5.13 is a pattern of period 2 called a *blinker* in ESPCA-01wxyf.
Figure 5.14 is a stable pattern called a *tub* in ESPCA-01wxyf.

As we shall see in Chap. 6, a rotor and a blinker are useful patterns in
ESPCAs 01c5ef and 01caef for constructing reversible Turing machines.

Note that in ESPCA-04wxyf *mirror images* of a rotor, a blinker and
a tub act as periodic patterns of period 4, 2 and 1, since corresponding
instances of ESPCA-04wxyf and ESPCA-01wxyf are dual under reflection
(see Definition 3.1). For example, ESPCA-01caef and ESPCA-04cabf are
dual under reflection, and thus they are essentially the same. Therefore,
we do not consider ESPCA-04wxyf.

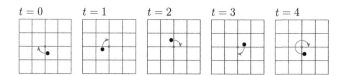

Fig. 5.12   *Rotor*, a periodic pattern of period 4 in ESPCA-01wxyf.

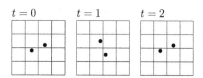

Fig. 5.13   *Blinker*, a periodic pattern of period 2 in ESPCA-01wxyf.

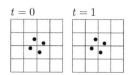

Fig. 5.14   *Tub*, a stable pattern in ESPCA-01wxyf.

Generally, in reversible and conservative ESPCAs, a large number of periodic or space-moving patterns exist, and their periods can be very long. This is because the number of particle is constant, and thus the same patterns must appear at the same or translated positions. For example, the patterns shown in Fig. 5.16 have very long periods in ESPCA-01c5ef (Fig. 5.15), which is an instance of ESPCA-01wxyf.

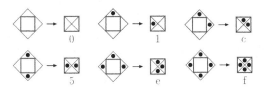

Fig. 5.15   Local function of reversible and conservative ESPCA-01c5ef.

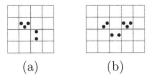

(a)   (b)

Fig. 5.16   Periodic patterns in ESPCA-01c5ef whose periods are (a) 304, and (b) 12288.

Note that a periodic pattern having arbitrarily long period can be obtained in several reversible and conservative ESPCAs by the method of using space-moving patterns and reflectors, which was explained in the case of GoL (Sec. 5.1.2.3). Such a construction of periodic patterns is possible, for example, in ESPCA-01c5ef (see Sec. 6.1.2). However, it is not known whether there are periodic patterns having arbitrarily long periods that use no space-moving pattern, in ESPCA-01wxyf.

ESPCA-08wxyf   Next, consider ESPCA-08wxyf (Fig. 5.17), where $w \in \{3, 6, 9, c\}$, $x \in \{5, a\}$, and $y \in \{7, b, d, e\}$. We can see that finite patterns in it are *all* periodic. The reason is as follows. A *bounding box* of a pattern is the smallest rectangular region that covers the pattern. If a particle goes out from the bounding box at some time, then the rule applied to the particle must be the second local transition rule of Fig. 5.17. By this rule, however, the particle is brought back to the region at the next step.

Fig. 5.17   Partial local function of ESPCA-08wxyf.

Therefore, no particle contained in the initial pattern can go out from the region by more than one cell. Therefore, neither space-moving pattern, nor diameter-growing pattern exists, and all the patterns become periodic. Thus, ESPCA-08wxyf is not interesting to watch its evolution processes.

5.2.1.2   *The case of reversible and non-conservative ESPCAs*

In reversible and non-conservative ESPCAs, periodic patterns are much fewer than the case of reversible and conservative ESPCAs. This is because many configurations of these ESPCAs become population-growing ones having no periodic pattern. However, some of these ESPCAs still have periodic patterns with interesting features.

ESPCA-01753f   First consider ESPCA-01753f (Fig. 5.18). A rotor, a blinker and a tub (Figs. 5.12–5.14) also exists in it. Figure 5.19 is another example of a periodic pattern. Since ESPCA-01753f is non-conservative, populations of the patterns sometimes increase and sometimes decrease. Periodic patterns having very long period also exist as shown in Fig. 5.20.

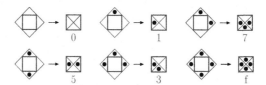

Fig. 5.18   Local function of reversible and non-conservative ESPCA-01735f.

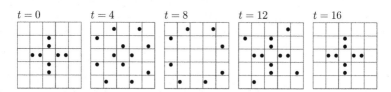

Fig. 5.19   Periodic pattern in non-conservative ESPCA-01753f whose period is 16.

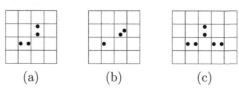

Fig. 5.20   Periodic patterns in ESPCA-01753f whose periods are (a) 760, (b) 2228, and (c) 73084.

ESPCA-0945df Next, consider the reversible and non-conservative ESPCA-0945df whose local function is given in Fig. 5.21.

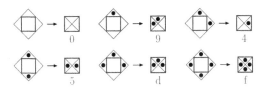

Fig. 5.21 Local function of reversible and non-conservative ESPCA-0945df.

There are various periodic patterns in ESPCA-0945df. Figures 5.22 and 5.23 show patterns of period 6 and period 22.

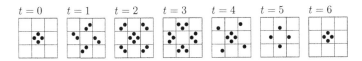

Fig. 5.22 Periodic pattern in ESPCA-0945df whose period is 6.

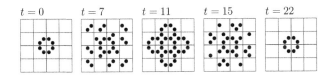

Fig. 5.23 Periodic pattern in ESPCA-0945df whose period is 22.

The pattern in Fig. 5.24, whose period is 60, has an interesting feature.

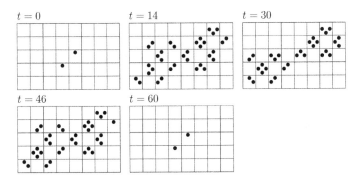

Fig. 5.24 Periodic pattern in ESPCA-0945df whose period is 60. The patterns at $t = 14$ and 46 have the maximum population 34, and those at $t = 14$, 30 and 46 have the maximum diameter 8.

Let p_{max} (p_{min}, respectively) be the maximum (minimum) population of the periodic pattern. Then, the ratio is $p_{max}/p_{min} = 17$, which is the largest value among the periodic patterns so far found. At first it grows in its population, but finally it shrinks and goes back to the two-particle pattern. It is not known whether there is a periodic pattern that has a ratio larger than 17, even in other ESPCAs. It is also not known whether there is an upper bound on the ratio.

5.2.2 *Periodic patterns in ETPCAs*

There are 36 reversible ETPCAs in total (see Fig. 3.20). If we consider only the ones with the quiescent state $(0, 0, 0)$, and exclude the dual ones under reflection, then there are ten ETPCAs in total. They are ETPCAs 0137, 0157, 0167, 0237 and 0257, which are conservative, and ETPCAs 0317, 0327, 0347, 0517 and 0527, which are non-conservative. Among them, ETPCAs 0157, 0137 and 0347 are important, since they will be shown to be Turing universal in Chap. 7.

5.2.2.1 *The case of reversible and conservative ETPCAs*

First, consider ETPCA-0237 (Fig. 5.25).

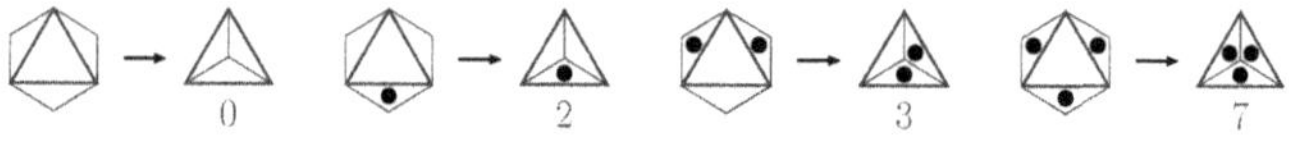

Fig. 5.25   Local function of ETPCA-0237.

In ETPCA-0237, every pattern is periodic by a similar reason as in ESPCA-08wxyf (Sec. 5.2.1.1). Namely, no particle can go out from the bounding box of the initial pattern by more than one cell.

It is also the case in ETPCA-0257. As shown in Fig. 3.21, every particle goes back and forth between a pair of adjacent cells, and thus every pattern has a period 2 or 1.

ETPCA-0167 (Fig. 5.26) has a similar property. There, every pattern is periodic with a period 6 or its factor. This is because we can interpret its local function as the one that makes every particle turn by $60°$.

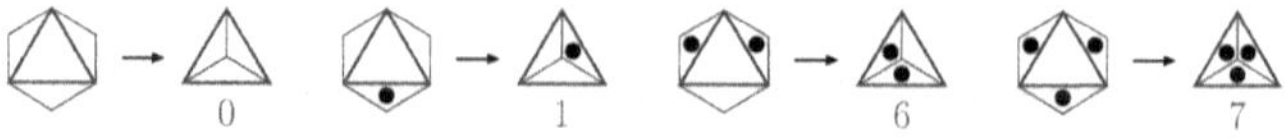

Fig. 5.26   Local function of ETPCA-0167.

In the above ETPCAs, no space-moving pattern exists. In the following, we investigate ETPCAs 0157 and 0137 in some detail.

ETPCA-0157   The local function of ETPCA-0157 is in Fig. 5.27.

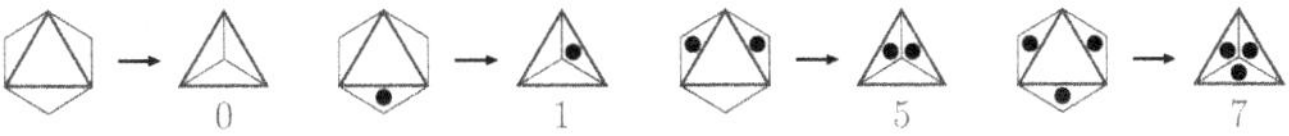

Fig. 5.27   Local function of ETPCA-0157.

In ETPCA-0157, a *light block* and a *heavy block* exist as stable patterns (Fig. 5.28). Since a heavy block will be shown to be useful in ETPCA-0157, it is simply called a *block* hereafter.

(a)   (b)

Fig. 5.28   (a) *Light block* and (b) *heavy block*, which are stable patterns in ETPCA-0157. In ESPCA-0157, a heavy block is simply called a *block* hereafter.

Figure 5.29 shows a periodic pattern of period 6 called a *rotor*. As we shall see later, it will be used to control a signal.

$t = 0$ $t = 1$ $t = 2$ $t = 3$ $t = 4$ $t = 5$ $t = 6$

Fig. 5.29   *Rotor*, a periodic pattern of period 6 in ETPCA-0157.

Similar to the case of ESPCAs, there are many periodic patterns having very long periods. Examples of such periodic patterns are given in Fig. 5.30.

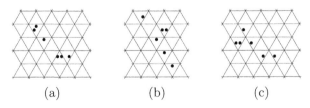

(a)   (b)   (c)

Fig. 5.30   Periodic patterns in ETPCA-0157. Their periods are (a) 13734, (b) 24066, and (c) 76158.

Interestingly, the patterns (b) and (c) of Fig. 5.30 move along a circular path in the cellular space like space-moving patterns. Figure 5.31 shows how the pattern Fig. 5.30(b) wander around. These patterns with very long periods are difficult to manage. Hence, we have not yet found their good applications so far.

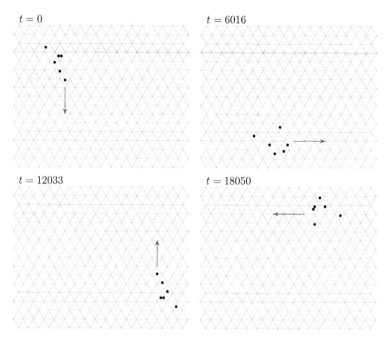

Fig. 5.31   Wandering process of the periodic pattern of Fig. 5.30(b) in ETPCA-0157. At $t = 24066$, it becomes the same configuration as the one at $t = 0$.

ETPCA-0137   Next, consider ETPCA-0137 (Fig. 5.32).

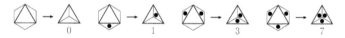

Fig. 5.32   Local function of ETPCA-0137.

In ETPCA-0137, a *light block* is stable (Fig. 5.33). On the other hand, a heavy block in ETPCA-0157 (Fig. 5.28(b)) is a periodic pattern of period 7 in ESPCA-0137. Hence, in ETPCA-0137, a light block is simply called a *block*. Later, a signal transmission wire will be composed out of blocks.

Fig. 5.33   *Light block*, a stable pattern in ETPCA-0137. In ESPCA-0137, a light block is simply called a *block*.

A *rotor* (Fig. 5.29) is also a periodic pattern in ETPCA-0137. It will be used to control a signal when composing reversible Turing machines.

In ETPCA-0137, many periodic patterns have long periods (Fig. 5.34) as in ETPCA-0157. Among them the pattern of Fig. 5.34(d) moves along a circular path, which is similar to the ones of Fig. 5.30(b) and (c).

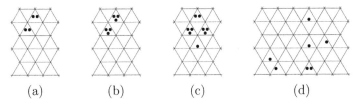

Fig. 5.34   Examples of periodic patterns in ETPCA-0137. Their periods are (a) 54, (b) 216, (c) 1410, and (d) 192426.

5.2.2.2   *The case of reversible and non-conservative ETPCAs*

In the five reversible and non-conservative ETPCAs 0317, 0327, 0347, 0517 and 0527, only ETPCA-0347 has various interesting periodic patterns. In the other four ETPCAs, no periodic patterns has been found so far.

ETPCA-0347   Here, consider ETPCA-0347 (Fig. 5.35). It is reversible and non-conservative.

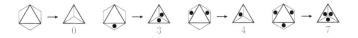

Fig. 5.35   Local function of ETPCA-0347.

In ETPCA-0347, a *medium-weight block* (Fig. 5.36) having nine particles is a stable pattern. It will be used to control the moving direction of a space-moving pattern. In ETPCA-0347, it is simply called a *block*.

Fig. 5.36   *Medium-weight block,* a stable pattern. It is simply called a *block* in ETPCA-0347.

Figure 5.37 shows several examples of periodic patterns in ETPCA-0347. In them, a *fin* and a *rotator* will be used to compose reversible Turing machines.

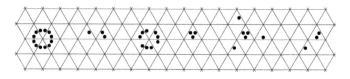

Fig. 5.37   Periodic patterns in ETPCA-0347. They are a *flashing light* (2), a *fin* (6), a *reflected block* (8), a *propeller* (8), a *large propeller* (24), and a *rotator* (42) (from left to right). The integers in the parentheses are their periods.

5.3   Space-Moving Patterns in Reversible PCAs

As in the case of GoL, space-moving patterns are important objects also in reversible PCAs. They can move from place to place, and interact with other patterns. We can use a space-moving pattern as a signal for constructing reversible computers, if its move direction and its phase are controlled by other patterns.

5.3.1   *Space-moving patterns in ESPCAs*

Space-moving patterns are found in many reversible ESPCAs. In the following, we explain how they behave in several selected ESPCAs.

5.3.1.1   *The case of reversible and conservative ESPCAs*

ESPCA-02wxyf   First, consider reversible and conservative ESPCA-02wxyf where $w \in \{3, 6, 9, c\}, x \in \{5, a\}$, and $y \in \{7, b, d, e\}$. Figure 5.38 shows three local transition rules that are common to all the 32 ESPCAs.

Fig. 5.38   Partial local function of ESPCA-02wxyf.

In ESPCA-02wxyf, a single particle works as a space-moving pattern of period 1 (Fig. 5.39). It is called a *glider-1*, whose speed is c.

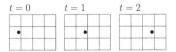

Fig. 5.39   *Glider-1*, a space-moving pattern of period 1 in ESPCA-02wxyf.

Any collection of glider-1's that move to the same direction is, obviously, a space-moving pattern having the speed of c. We can see that no space-moving pattern exists other than that in ESPCA-02wxyf as shown below.

Assume there is a space moving pattern of period m whose displacement vector in one period is (d_x, d_y). Let n be the total number of particles in the pattern. Since its speed does not exceed the speed of light, $d_x + d_y \leq m$ holds. Without loss of generality, we assume $d_x > 0$ and $d_y \geq 0$ (otherwise, rotate the pattern appropriately).

First, consider the case where $d_x = m$ (thus $d_y = 0$). Then, the total distance that the n particles move in one period is nm. Therefore, each particle must move rightward by one cell every step. If the second local transition rule in Fig. 5.38 is applied to some particle for shifting it upward, downward or leftward, then it does not satisfy the condition. Likewise, if some local transition rule of ESPCA-02wxyf corresponding to w, x, y or f, which shifts two or more particles in the neighbor cells to the center cell, is applied to some cell of the pattern, then the horizontal component of the displacement vector of one of the particles must be 0 or -1 in this step. Hence, it is not possible to apply such a rule. Thus, there is no choice other than to use the second local transition rule for shifting each particle to the right. Therefore, in this case, the pattern must be a collection of glider-1's that move rightward.

Second, consider the case where $0 < d_x < m$. Let (x_R, y_R) be the coordinates of one of the rightmost non-blank cells of the pattern. Since $d_x > 0$, there exists y_0 such that the cell $(x_R + 1, y_0)$ first becomes a non-blank state at time t_0. Since its neighbor cells except the left neighbor are in the blank state at $t_0 - 1$, the state of the cell $(x_R + 1, y_0)$ must be $(0, 1, 0, 0)$ (*i.e.*, one particle is in the right part) by the second local transition rule. Then, at time $t_0 + 1$, the cell $(x_R + 2, y_0)$ also becomes the state $(0, 1, 0, 0)$, and so on. Therefore, at least one particle moves rightward with the speed of light after t_0, *i.e.*, it has the displacement vector $(m, 0)$ in one period. It contradicts the assumption that the whole pattern has the displacement vector (d_x, d_y) with $d_x < m$, and thus the second case is impossible.

As we shall see in Secs. 6.3 and 6.4, the pattern composed of two consecutive glider-1's, which is called a *glider-1w* (Fig. 6.30), is used as a signal in ESPCAs 02c5df and 02c5bf for constructing reversible Turing machines.

ESPCA-01c*xy*f   Second, consider reversible and conservative ESPCA-01c*xy*f where $x \in \{5, a\}$, and $y \in \{7, b, d, e\}$ (Fig. 5.40).

Fig. 5.40   Partial local function of ESPCA-01c*xy*f.

In ESPCA-01c*xy*f, a space-moving pattern of period 12 called a *glider-12* exists (Fig. 5.41). Its speed is $c/6$. As it will be seen in Secs. 6.1 and 6.2, it used as a signal for composing reversible Turing machines in ESPCAs 01c5ef and 01caef.

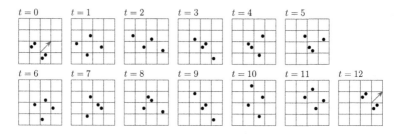

Fig. 5.41   *Glider-12*, a space-moving pattern of period 12 in ESPCA–01c*xy*f.

In each instance of ESPCA-01c*xy*f, many kinds of space-moving patterns exist. Figure 5.42 shows six examples of space-moving patterns in ESPCA-01caef, an instance of ESPCA-01c*xy*f. Since the periods of them except the glider-12 are large, it is difficult to find a method for controlling them. Hence, so far, no good application of them has been found.

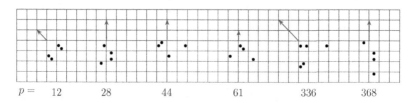

Fig. 5.42   Various examples of space-moving patterns in ESPCA-01caef. Their periods are indicated below. Arrows represent displacement vectors in one period.

ESPCA-016a7f   Next, consider reversible and conservative ESPCA-016a7f (Fig. 5.43).

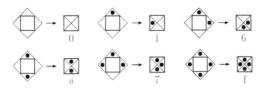

Fig. 5.43   Local function of ESPCA-016a7f.

Figure 5.44 shows a space-moving pattern of period 3 in ESPCA-016a7f. Its speed is $c/3$. It exists also in ESPCA-01657f. Although its period is short, no method of using it as a signal has been found. Figure 5.45 gives various examples of space-moving patterns in ESPCA-016a7f. Among them, the pattern of period 10 exists also in ESPCA-016ayf where $y \in \{7, b, d, e\}$.

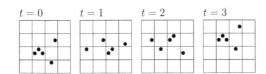

Fig. 5.44   Space-moving pattern of period 3 in ESPCA-016a7f.

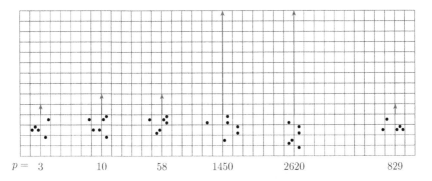

Fig. 5.45   Examples of space-moving patterns in ESPCA-016a7f. Their periods are indicated below. Arrows represent displacement vectors in one period.

5.3.1.2 *The case of reversible and non-conservative ESPCAs*

So far, only several kinds of space-moving patterns have been found in non-conservative ESPCAs. We list all of them below. Some of them have short periods. However, no method of using them as signals has been found so far. If we interact such a pattern with another pattern, an *explosion*, a rapid increase of population, occurs very often, since these ESPCAs are non-conservative. Thus, it is difficult to control them.

ESPCA-094xyf   Consider ESPCA-094xyf where $x \in \{5, a\}$, and $y \in \{7, b, d, e\}$ (Fig. 5.46). A space-moving pattern of period 3 called a *glider-3* (Fig. 5.47) exists in it.

Fig. 5.46   Partial local function of ESPCA-094xyf.

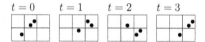

Fig. 5.47   *Glider-3*, a space-moving pattern of period 3 in ESPCA-094xyf.

ESPCA-098a7f   Consider ESPCA-098a7f (Fig. 5.48). There are two kinds of space-moving patterns in ESPCA-098a7f. They are a *glider-5* (Fig. 5.49), and a *glider-15* (Fig. 5.50).

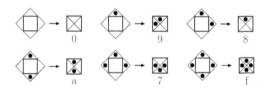

Fig. 5.48   Local function of ESPCA-098a7f.

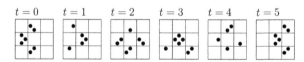

Fig. 5.49   *Glider-5*, a space-moving pattern of period 5 in ESPCA-098a7f.

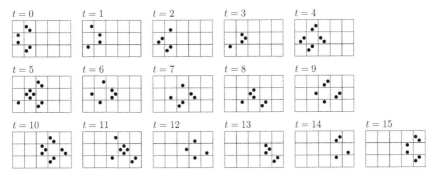

Fig. 5.50   *Glider-15*, a space-moving pattern of period 15 in ESPCA-098a7f.

ESPCA-098$x$df   Consider ESPCA-098$x$df where $x \in \{5, a\}$ (Fig. 5.51). A space-moving pattern of period 5 called a *glider-5a* (Fig. 5.52) exists.

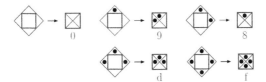

Fig. 5.51   Partial local function of ESPCA-098$x$df.

Fig. 5.52   *Glider-5a*, a space-moving pattern of period 5 in ESPCA-098$x$df.

ESPCA-098aef   Consider ESPCA-098aef (Fig. 5.53). A space-moving pattern of period 10 called a *glider-10* (Fig. 5.54) exists in it. Note that the pattern at $t = 0$ ($t = 5$, respectively) in Fig. 5.54 is the same as that in Fig. 5.49 (Fig. 5.50), their evolution processes are different.

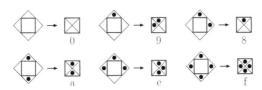

Fig. 5.53   Local function of ESPCA-098aef.

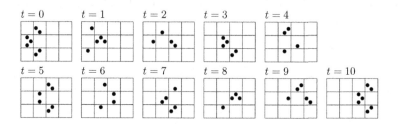

Fig. 5.54   *Glider-10*, a space-moving pattern of period 10 in ESPCA-098aef.

ESPCA-0c4a7f   Consider ESPCA-0c4a7f (Fig. 5.55). A space-moving pattern of period 15 called a *glider-15a* (Fig. 5.56) exists in it.

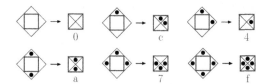

Fig. 5.55   Local function of ESPCA-0c4a7f.

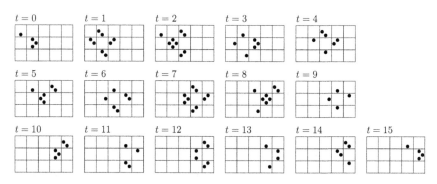

Fig. 5.56   *Glider-15a* a space-moving pattern of period 15 in ESPCA-0c4a7f.

5.3.2   *Space-moving patterns in ETPCAs*

We show examples of space-moving patterns in some ETPCAs. A glider-6 in ETPCA-0347 exhibits interesting behavior. However, no other useful space-moving pattern has been found so far.

5.3.2.1 The case of reversible and conservative ETPCAs

As discussed in Sec. 5.2.2.1, there is no space-moving pattern in ETPCAs 0167, 0237 and 0257. Those in ESPCAs 0157 and 0137 are shown below.

ETPCA-0157 Consider ETPCA-0157 (Fig. 5.27). There are many space-moving patterns in it. Figure 5.58 shows three examples. However, since their periods large, it is difficult to control them. It is not known whether there exist ones with short periods. Thus, in Sec. 7.3, reversible Turing machines will be constructed without using a space-moving pattern.

$p =$ 1016   1576   2854

Fig. 5.57   Examples of space-moving patterns in ETPCA-0157. Their periods are indicated below. Arrows represent displacement vectors in one period.

ETPCA-0137 Next, consider ETPCA-0137 (Fig. 5.32). Many space-moving pattern having long periods also exist in it. Figure 5.58 shows three examples. Again, in Sec. 7.4, reversible Turing machines will be constructed without using a space-moving pattern.

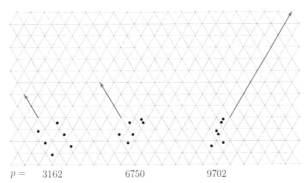

$p =$ 3162   6750   9702

Fig. 5.58   Examples of space-moving patterns in ETPCA-0137. Their periods are indicated below. Arrows represent displacement vectors in one period.

5.3.2.2 *The case of reversible and non-conservative ETPCAs*

In reversible and non-conservative ETPCAs, no space-moving pattern has been found except in ETPCA-0347.

ETPCA-0347 Consider ETPCA-0347 (Fig. 5.35). A space-moving pattern of period 6 called a *glider-6* exists in it (Fig. 5.59). It swims in the cellular space like a fish or an eel. As we shall see below, it is a particularly interesting and useful pattern. Interacting with another glider-6 or other patterns, various fantastic phenomena appear. Using a glider-6 as a signal, Turing machines will be constructed in Sec. 7.1.

On the other hand, it is not known whether there exists another space-moving pattern that is essentially different from a glider-6.

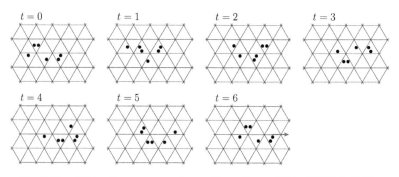

Fig. 5.59   *Glider-6*, a space moving pattern of period 6 in ETPCA-0347.

5.4   Interactions of Patterns in Reversible PCAs

We investigate how space-moving patterns interact with periodic ones or other space-moving ones in reversible PCAs, and how such interactions can be used for composing functional modules.

5.4.1   *Interactions of space-moving and periodic patterns*

To use a space-moving pattern (or a glider) as a signal, it is necessary to control its movements by interacting with periodic patterns (preferably by stable patterns). In GoL, the moving direction of a glider is changed by various periodic patterns, like a snark (Fig. 5.10). In such a processes, neither the position nor the phase of the periodic pattern should be changed. Otherwise, a difficulty arises in reusing the control mechanism.

It is also convenient if a position of a small periodic pattern can be shifted by interacting a space-moving pattern with it. By this, it becomes possible to use the periodic pattern as a memory device where the memory states are distinguished by the position of the periodic pattern. In GoL, a block can be shifted by colliding gliders with it (Fig. 5.9).

These operations are performed in several reversible elementary PCAs. We explain them using some selected PCAs below.

ETPCA-0347   Consider ETPCA-0347 (Fig. 5.35). In this cellular space, the moving direction of a glider-6 (Fig. 5.59) is changed by a pattern composed of blocks. Figure 5.60 shows a process of a 120°-right-turn of a glider-6 by a sequence of three blocks.

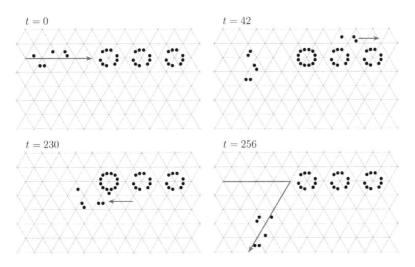

Fig. 5.60   Colliding a glider-6 with a sequence of three blocks in ETPCA-0347, the glider-6 makes a 120°-right-turn.

As seen in the figure at $t = 42$, if the glider-6 collides with the blocks, it first decomposed into a body (left), which is a periodic pattern called a rotator (Fig. 5.37), and a fin (upper right). Here, the fin moves around the sequence of blocks twice clockwise (note that the fin itself is a periodic pattern (Fig. 5.37)). Finally, the fin meets the body at $t = 230$. There, the fin is attached to the body to reconstruct a glider-6. Then, the glider-6 moves to the south-west direction ($t = 256$). By this, a *120°-right-turn gadget* is obtained.

By changing the arrangement of blocks, we have several kinds of turn gadgets, such as the ones of 60°-right-turn, backward-turn and U-turn. They are explained in Sec. 7.1.2. Combining these turn gadgets, we can also adjust the phase of a glider-6 (see also Sec. 7.1.2).

Interacting a glider-6 with a fin appropriately, the position of the fin is shifted. Figure 5.61 is a pushing process of a fin. It is also possible to pull a fin by a glider-6 (see Fig. 7.10). Therefore, a fin is usable as a *position marker* to realize a memory device. In Sec. 7.1.3, using these phenomena, a composing method of a reversible logic element with memory (RLEM) will be given. By this method, rather than the method of using only reversible logic gates, construction of reversible Turing machines is greatly simplified, since there is no need of synchronizing two or more signals.

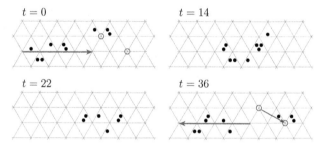

Fig. 5.61   Interaction of a glider-6 and a fin in ETPCA-0347. By this, the fin is pushed, and the glider-6 goes back along the same path.

ESPCA-01c5ef   Consider ESPCA-01c5ef (Fig. 5.15). In ESPCA-01c5ef, a glider-12 exists (Fig. 5.41). Colliding a glider-12 with a rotor as in Fig. 5.62, it makes a left-turn. In this way, a glider-12 can work as a signal.

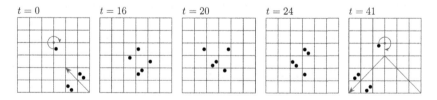

Fig. 5.62   Interaction of a glider-12 and a rotor in ESPCA-01c5ef. The glider-12 makes a left-turn.

Colliding a glider-12 with a rotor in another way as shown in Fig. 5.63, the rotor is shifted by four cells. Thus, a rotor can also be used as a position marker. In Sec. 6.1, this process will be used for composing RLEMs.

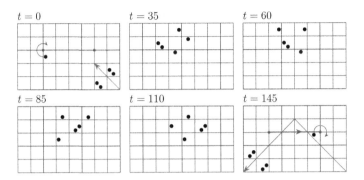

Fig. 5.63   Interaction of a glider-12 and a rotor in ESPCA-01c5ef. In this case, the rotor is shifted right.

ESPCA-01caef   Next, consider ESPCA-01caef (Fig. 5.64).

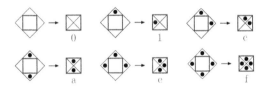

Fig. 5.64   Local function of ESPCA-01caef.

The same glider-12 (Fig. 5.41) as in ESPCA-01c5ef exists also in ESPCA-01caef. In this case, colliding a glider-12 with a blinker by the method shown in Fig. 5.65, it makes a right-turn.

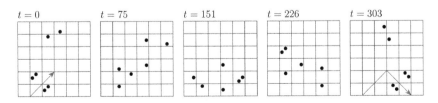

Fig. 5.65   Interaction of a glider-12 and a blinker in ESPCS-01caef. The glider-12 makes a right-turn.

On the other hand, colliding a glider-12 with a blinker as shown in Fig. 5.66, the blinker is shifted by six cells. Thus, a blinker can be used as a position marker. In Sec. 6.2.3, it is used for composing RLEMs. Note that the shifting process takes more than 2000 steps, and the particles in the pattern show very complex behavior during this process.

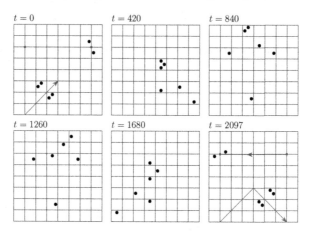

Fig. 5.66   Interaction of a glider-12 and a blinker in ESPCS-01caef. Here, the blinker is shifted left.

5.4.2   *Interactions of two or three space-moving patterns*

We show some interesting phenomena obtained by interacting two or three space-moving patterns.

ETPCA-0347   First, consider ETPCA-0347 (Fig. 5.35). In this reversible and non-conservative ETPCA, glider-6's show interesting behavior.

Colliding two glider-6's as in Fig. 5.67, the glider-6 coming from the port x changes its moving direction, while the other from c goes straight ahead as if nothing happened. This process simulates a kind of reversible logic gate called a *switch gate* [52], which realizes the 2-input 3-output logic function $(c, x) \mapsto (c, cx, \bar{c}x)$. It is known that a Fredkin gate, which is a 3-input 3-output universal reversible gate, is constructed using switch gates and inverse switch gates. Therefore, it is in principle possible to construct reversible Turing machines using only these gates. However, if we do so, timing of glider-6's should be adjusted exactly at each gate, and thus the design of the configuration becomes complicated. In Sec. 7.1, a method of constructing an RLEM without using a logic gate is presented.

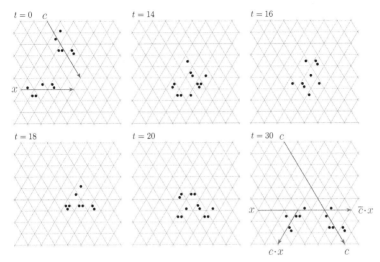

Fig. 5.67   Interacting two glider-6's in ETPCA-0347, a switch gate is implemented [9].

In GoL, by colliding two gliders appropriately, one of them or both of them can be erased, as seen in Sec. 5.1.2.1. Since such phenomena are irreversible processes, they cannot be seen in a reversible PCA. Instead, increasing the number of glider-6's from two to three, and decreasing it from three to two are possible in ETPCA-0347. These processes are shown in Figs. 5.68 and 5.69. Note that they are time-symmetric processes each other as discussed in Example 3.9.

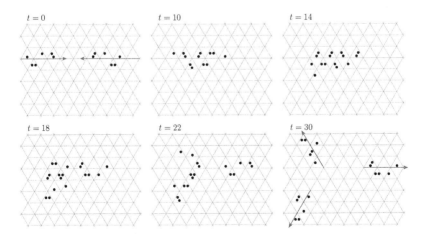

Fig. 5.68   Interacting two glider-6's in ETPCA-0347, three are generated.

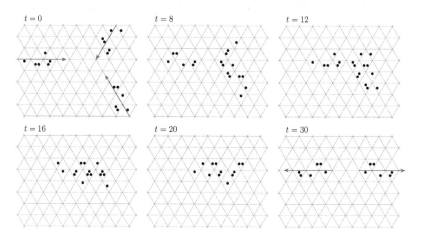

Fig. 5.69   Interacting three glider-6's in ETPCA-0347, one of them is reversibly erased.

ESPCA-02c5yf   Next, consider ESPCA-02c5yf where $y \in \{7, b, d, e\}$ (Fig. 5.70).

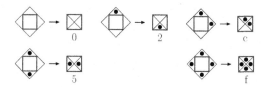

Fig. 5.70   Partial local function of ESPCA-02c5yf.

In ESPCA-02$wxyf$ (hence in ESPCA-02c5yf), a single particle called a glider-1 works as a space-moving pattern of period 1, as seen in Fig. 5.39. On the other hand, no position marker, which is a periodic pattern that can be shifted by a space-moving pattern, has been found in ESPCA-02c5yf. Therefore, in Secs. 6.3.2 and 6.4.2, interaction of two signals is used as an elementary operation. There, a space-moving pattern composed of two glider-1's called a *glider-1w* is used as a signal (Fig. 5.71).

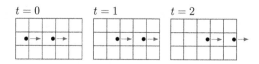

Fig. 5.71   *Glider-1w* composed of two glider-1's is used as a signal in ESPCA-02c5yf.

Colliding two glider-1w's as in Fig. 5.72, a kind of a reversible logic gate called an *interaction gate* [52] is realized. It is a 2-input 4-output gate that implements the logical function $(x_1, x_2) \mapsto (x_1 x_2, \overline{x_1} x_2, x_1 \overline{x_2}, x_1 x_2)$. In Secs. 6.3.3 and 6.4.2, a rotary element (RE) will be composed out of interaction gates and inverse interaction gates.

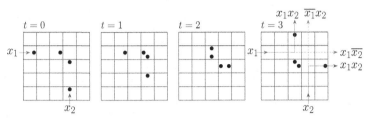

Fig. 5.72   Interacting two glider-1w's in ESPCA-02c5yf, an interaction gate is realized.

5.5   Glider Guns in Reversible PCAs

A *glider gun* is a pattern that periodically generates space-moving patterns. Hence, it is a population-growing one. As we have already seen, there is a famous glider gun by Gosper (Fig. 5.6) in GoL. Below, we can see several glider guns exist also in reversible PCAs. Note that these PCAs must be non-conservative.

Here, we do not give a formal definition of a gun, since a precise definition will become cumbersome in its details. Instead, we give an intuitive explanation using an example.

5.5.1   *Glider guns in ESPCAs*

We give three examples of glider guns in ESPCAs.

ESPCA-094x7f   Consider ESPCA-094x7f where $x \in \{5, a\}$ (Fig. 5.73). It is a subcase of ESPCA-094xyf (Fig. 5.46), and thus a glider-3 (Fig. 5.47) exists in it.

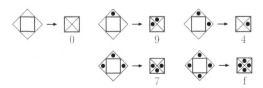

Fig. 5.73   Partial local function of ESPCA-094x7f.

Figure 5.74 shows a glider gun in ESPCA-094x7f. It generates four glider-3's every 8 steps. Compare, for example, the configurations at $t = 4$ and $t = 12$. We can see four glider-3's are newly created at $t = 12$, and they move away. Since the pattern in the box at $t = 12$ is the same as the one at $t = 4$, four more glider-3's will be generated at $t = 20$, and so on. Thus, the configurations in this figure generate glider-3's indefinitely

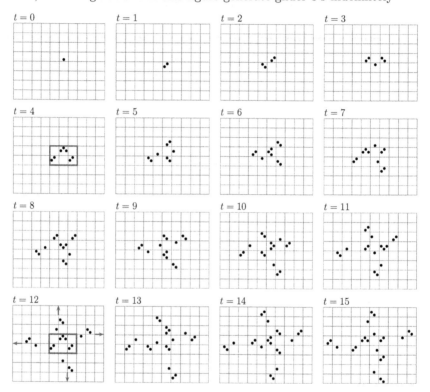

Fig. 5.74   Four-way glider gun in ESPCA-094x7f. Starting from a single particle ($t = 0$), it generates four glider-3's every 8 steps.

How will the configuration of $t = 0$ evolve if we go back to the past? Interestingly, it acts as a glider gun also to the negative time direction as shown in Fig. 5.75. In other words, from $t = -60$ four gliders are absorbed every 8 steps until the configuration becomes a single particle at $t = 0$.

It is easy to see that ESPCAs 09457f and 094a7f, which are instances of ESPCA-094x7f, are both T-symmetric under H^{rev}. This is because their local functions f_{09457f} and f_{094a7f} satisfy $f_{09457f}^{-1} = f_{09457f}$ and $f_{094a7f}^{-1} = f_{094a7f}$ (Theorem 3.1). Let F be the global function of ESPCA-09457f or

ESPCA-094a7f. Let $\alpha(t)$ denote the configuration of ESPCA-094x7f at t, where $\alpha(0)$ is the single-particle configuration. Then, by Lemma 3.5 the following relation holds for any $t > 0$.

$$\alpha(-t) = (F^{-1})^t(\alpha(0)) = H^{\text{rev}} \circ F^t \circ H^{\text{rev}}(\alpha(0))$$

Since $\alpha(0)$ is a single-particle configuration, $H^{\text{rev}}(\alpha(0))$ can be regarded as the 180°-rotated configuration of $\alpha(0)$. Hence, $F^t \circ H^{\text{rev}}(\alpha(0))$ is the 180°-rotated configuration of $F^t(\alpha(0)) = \alpha(t)$. Therefore, $\alpha(-t)$ is the 180°-rotated configuration of $H^{\text{rev}}(\alpha(t))$, which is very similar to $\alpha(t)$. In this sense, the evolution process is symmetric with respect to $t = 0$.

Fig. 5.75 The glider gun in ESPCA-094x7f generates glider-3's both to the positive and negative time directions. In other words, glider-3's are absorbed in $t < 0$, while they are generated in $t > 0$.

Behavior in the backward evolution depends on the local function and the configuration at $t = 0$. Other possibilities of backward evolution in glider guns will be given for ETPCA-0347 in Sec. 5.5.2.

ESPCA-094$x$df Consider ESPCA-094$x$df where $x \in \{5, a\}$ (Fig. 5.76). It is a subcase of ESPCA-094xyf (Fig. 5.46), and a glider-3 exists in it.

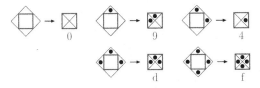

Fig. 5.76 Partial local function of ESPCA-094$x$df.

As shown in Fig. 2.11, there is a two-way glider gun in ESPCA-094$x$df. It generates two glider-3's every 10 steps. Similar to ESPCA-094x7f, the gun generates both to the forward and backward time directions (Fig. 5.77).

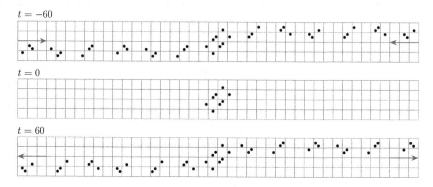

Fig. 5.77 Two-way glider gun in ESPCA-094$x$df. It generates glider-3's both to the positive and negative time directions.

ESPCA-098aef Consider ESPCA-098aef (Fig. 5.53). In this ESPCA, a glider-10 exists (Fig. 5.54). Figure 5.78 shows a four-way glider gun that generates four glider-10's every 18 steps.

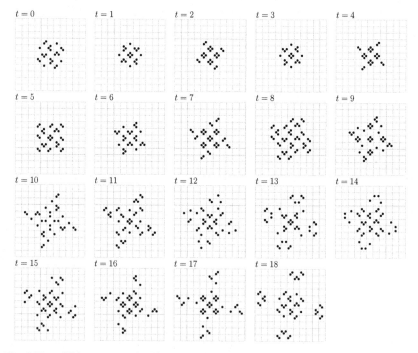

Fig. 5.78 Glider gun in ESPCA-098aef. It generates four glider-10's every 18 steps.

The glider gun in Fig. 5.78 generates also to the backward time direction as in Fig. 5.79.

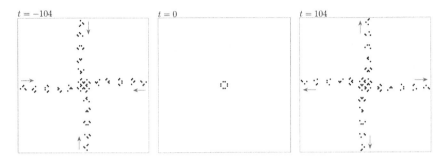

Fig. 5.79   The glider gun in ESPCA-098aef generates glider-10's both to the positive and negative time directions.

5.5.2   *Glider guns in ETPCAs*

In ETPCAs, glider guns have been found only in ETPCA-0347 (Fig. 5.35).

ETPCA-0347   Figure 5.80 shows that by colliding a glider-6 with a fin as shown at $t = 0$, a 3-way glider gun, which generates three glider-6's every 24 steps, is created.

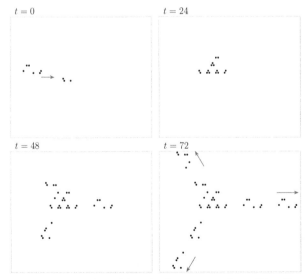

Fig. 5.80   3-way glider gun in ETPCA-0347. It generates three glider-6's every 24 steps.

If we go back to the negative time direction from $t = 0$ of Fig. 5.80, the position of the glider-6 moves far to the left, while that of the fin is at the same position. Therefore, unlike Fig. 5.79, the gun in Fig. 5.80 does not generate glider-6's to the backward time direction.

However, since ETPCA-0347 is T-symmetric under the involution $H^{\text{rev}} \circ H^{\text{refl}}$ (see Corollary. 3.2), a glider absorber is obtained by applying the involution to the configuration at $t = 72$ of Fig. 5.80. Combining the glider absorber and the glider gun, we have a configuration that generates gliders both in the forward and backward time directions as in Fig. 5.81.

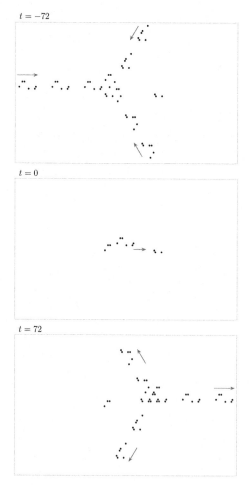

Fig. 5.81   Combining a glider absorber and a gun in ETPCA-0347, we have a gun that generates glider-6's both in the positive and negative time directions.

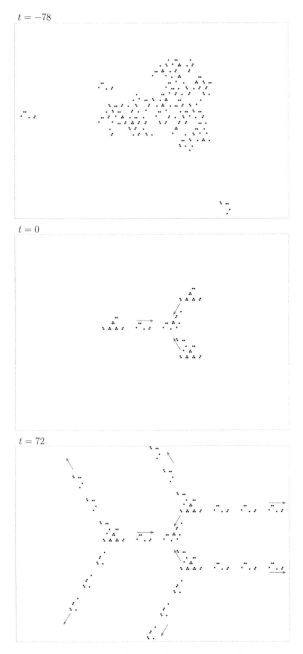

$t = -78$

$t = 0$

$t = 72$

Fig. 5.82   From a disordered pattern ($t = -78$), a 6-way gun that generates six glider-6's every 24 steps appears. It consists of three glider guns and one glider absorber.

Further combining three glider guns and one glider absorber as shown at $t = 72$ of Fig. 5.82, we have a 6-way glider gun that generates six glider-6's every 24 steps after $t = 0$. In this case, if we go back to the negative time direction, a disordered (*i.e.*, random-like) patterns appear.

There is yet another method of composing a glider gun in ETPCA-0347. This method uses the process of creating three glider-6's from two shown in Fig. 5.68. Using two glider-6's out of the generated three, three glider-6's are obtained again, and so on. A glider gun is constructed by controlling this process. The details of the method will be explained in Sec. 7.1.5.

5.6   Generating Fantastic Patterns

Non-conservative ESPCAs and ETPCAs often generate fantastic patterns even when they start from a simple pattern. In this section, we investigate pattern generation capabilities of linear and nonlinear ESPCAs, and linear ETPCAs. Note that several irreversible linear PCAs are contained in this section, since they give beautiful pictures such as fractal-like ones.

5.6.1   *Growing patterns in linear ESPCAs*

A *linear cellular automaton* is a CA whose local function is a linear function of the states of the neighboring cells [60]. More precisely, the set of states of a cell is $\mathbb{Z}_m = \{0, 1, \ldots, m - 1\}$ and the local function $f : \mathbb{Z}_m^n \to \mathbb{Z}_m$ is expressed as below for some $a_1, \ldots, a_n \in \mathbb{Z}_m$.

$$f(x_1, \ldots, x_n) = a_1 x_1 + \cdots + a_n x_n \quad (\bmod\ m)$$

We newly define a linear SPCA as follows.

Definition 5.1. Let P be an SPCA whose state set is $\mathbb{Z}_m^4$ (*i.e.*, each part has the state set $\mathbb{Z}_m$), and $f : \mathbb{Z}_m^4 \to \mathbb{Z}_m^4$ be its local function. P is called a *linear SPCA* if the following holds.

$$\exists\, a_1, a_2, a_3, a_4 \in \mathbb{Z}_m :$$
$$\forall\, t, r, b, l, t', r', b', l' \in \mathbb{Z}_m :$$
$$\text{if } f(t, r, b, l) = (t', r', b', l') \text{ then}$$
$$t' = a_1 t + a_2 r + a_3 b + a_4 l \quad (\bmod\ m)$$
$$r' = a_1 r + a_2 b + a_3 l + a_4 t \quad (\bmod\ m)$$
$$b' = a_1 b + a_2 l + a_3 t + a_4 r \quad (\bmod\ m)$$
$$l' = a_1 l + a_2 t + a_3 r + a_4 b \quad (\bmod\ m)$$

Note that, by the above definition, a linear SPCA is rotation-symmetric. Its proof is omitted here, since it is easily shown. Although it is possible to define a linear SPCA that may not be rotation-symmetric, here we consider only rotation-symmetric ones for simplicity.

Definition 5.2. A linear SPCA P is called a *linear ESPCA*, if its state set is $\{0,1\}^4$. In this case, the addition is $\oplus$, the mod 2 addition.

There are 16 linear ESPCAs, since each linear ESPCA is defined by $(a_1, a_2, a_3, a_4) \in \{0,1\}^4$. They are shown in Table 5.1. For example, if $(a_1, a_2, a_3, a_4) = (0,0,1,1)$, then the identification number $uvwxyz$ of the ESPCA is 095f60. This is because $(t', r', b', l') = (b \oplus l, l \oplus t, t \oplus r, r \oplus b)$ by Definition 5.1, and thus, *e.g.*, $f(0,0,1,0) = (1,0,0,1)$, which means $v = 9$.

Table 5.1   16 linear ESPCAs

(a_1, a_2, a_3, a_4)	ESPCA-$uvwxyz$	Reversible	Conservative
(0,0,0,0)	000000		
(0,0,0,1)	0195bf	✓	✓
(0,0,1,0)	08cadf	✓	✓
(0,0,1,1)	095f60		
(0,1,0,0)	0465ef	✓	✓
(0,1,0,1)	05f050		
(0,1,1,0)	0caf30		
(0,1,1,1)	0d3a8f	✓	
(1,0,0,0)	023a7f	✓	✓
(1,0,0,1)	03afc0		
(1,0,1,0)	0af0a0		
(1,0,1,1)	0b651f	✓	
(1,1,0,0)	065f90		
(1,1,0,1)	07ca2f	✓	
(1,1,1,0)	0e954f	✓	
(1,1,1,1)	0f00f0		

Local functions of the 16 linear ESPCAs are shown in Fig. 5.83. There are eight reversible ESPCAs in the 16 linear ones as noted in Table 5.1. Four of them are conservative. The following four pairs of ESPCAs are dual under reflection: (0195bf, 0465ef), (095f60, 0caf30), (03afc0, 065f90) and (0b651f, 0e954f). Since two ESPCAs in each pair are essentially the same, we consider only one of them in the following.

In what follows, we consider five linear irreversible ESPCAs 0f00f0, 05f050, 095f60, 0af0a0 and 03afc0, and two linear reversible ESPCAs 0d3a8f and 07ca2f. We investigate how they generate interesting patterns. Some of them show self-replication of patterns similar to the Fredkin CA, and some others generate fractal-like patterns.

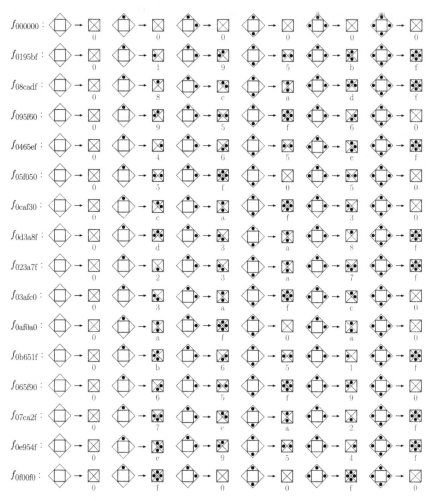

Fig. 5.83   Local functions of the 16 linear ESPCAs.

ESPCA-0f00f0   It is a linear irreversible ESPCA. Its local function f_{0f00f0} has the following property. If a cell receives an odd number of particles from the four neighboring cells, then the cell goes to the state $(1,1,1,1)$. Otherwise, it goes to the state $(0,0,0,0)$. Therefore, ESPCA-0f00f0 acts just like the Fredkin's self-replicating CA (see Example 1.1) if its initial pattern consists only of the states $(0,0,0,0)$ and $(1,1,1,1)$. Figure 5.84 shows a self-replicating process of a Cheshire-Cat-like pattern. Note that, in this section, the state 1 in a pattern is indicated by a small filled triangle rather than a dot.

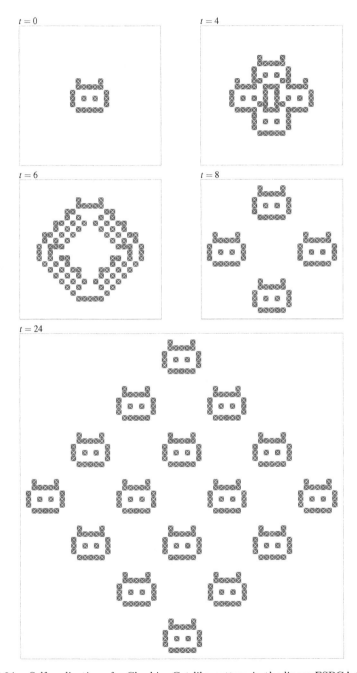

Fig. 5.84   Self-replication of a Cheshire-Cat-like pattern in the linear ESPCA-0f00f0.

ESPCA-05f050   It is a linear irreversible ESPCA. In this cellular space, any initial pattern consisting only of the states (0,0,0,0), (0,1,0,1), (1,0,1,0) and (1,1,1,1) generates its copies indefinitely as shown in Fig. 5.85. Its proof is left as an exercise for the readers (see Exercise 5.2). Note that an initial pattern (*i.e.* at $t = 0$) having a state other than the above does not self-replicate. However, the pattern at $t = 1$ satisfies the condition by its local function, and thus the pattern self-replicates.

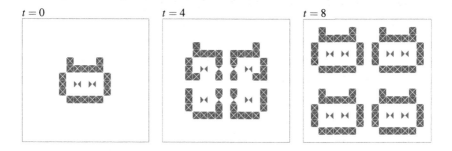

Fig. 5.85   Self-replication of a Cheshire-Cat-like pattern in the linear ESPCA-05f050.

ESPCA-095f60   It is a linear irreversible ESPCA. Here, any initial pattern consisting only of the states (0,0,0,0), (1,1,0,0), (0,1,1,0), (0,0,1,1), (1,0,0,1), (0,1,0,1), (1,0,1,0) and (1,1,1,1) creates its copies as shown in Fig. 5.86.

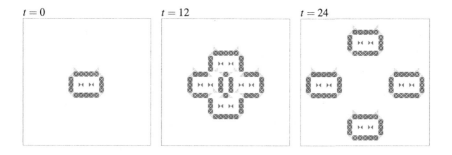

Fig. 5.86   Self-replication of a Cheshire-Cat-like pattern in the linear ESPCA-095f60.

ESPCA-0af0a0   It is a linear irreversible ESPCA. In this cellular space, a pattern consisting only of the state $(0,0,0,0)$ and $(1,0,1,0)$, and a pattern consisting only of $(0,0,0,0)$ and $(0,1,0,1)$ produce their copies independently as shown in Fig. 5.87. In the initial pattern, however, the state $(1,1,1,1)$, which is a superposition of $(1,0,1,0)$ and $(0,1,0,1)$, is allowed.

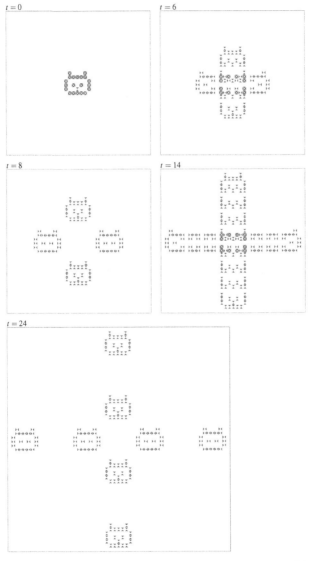

Fig. 5.87   Self-replication of a Cheshire-Cat-like pattern in the linear ESPCA-0af0a0.

ESPCA-03afc0   It is a linear irreversible ESPCA. The initial pattern (given at $t = 0$ of Fig. 5.88) is rotation-symmetric, and occupies both odd parity points and even parity points. Hence, a generated pattern at each time step is also rotation-symmetric and occupies both odd and even parity points. This ESPCA generates patterns that look like a carpet as shown in Fig. 5.88.

$t = 0$

$t = 48$

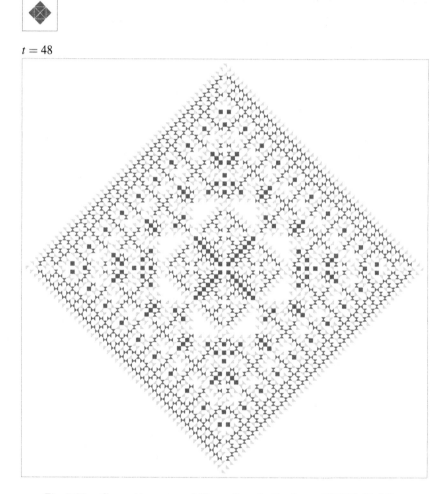

Fig. 5.88   Generating a carpet-like pattern in the linear ESPCA-03afc0.

ESPCA-0d3a8f   It is a linear reversible ESPCA. We assume the same initial pattern as in Fig. 5.88 ($t = 0$) is given. Then its evolution process is as in Fig. 5.89, and a fractal-like patterns are generated in it.

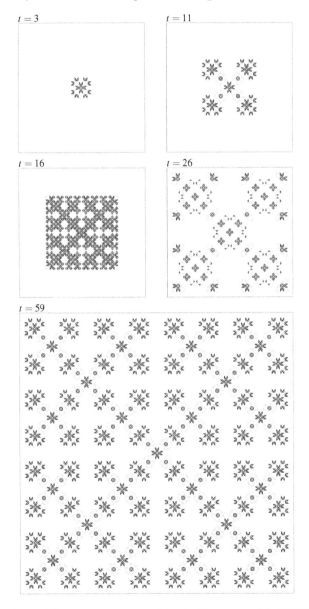

Fig. 5.89   Fractal-like patterns generated by the linear reversible ESPCA-0d3a8f.

ESPCA-07ca2f　　It is a linear reversible ESPCA. It generates fractal-like patterns shown in Fig. 5.90 for the initial pattern given in Fig. 5.88 ($t = 0$).

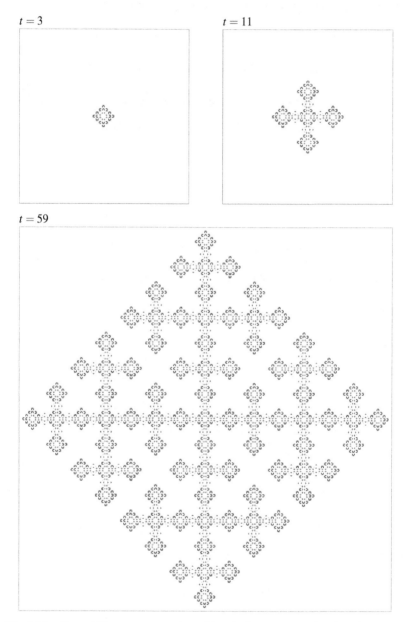

Fig. 5.90　　Fractal-like patterns generated by the linear reversible ESPCA-07ca2f.

5.6.2 *Growing patterns in non-linear reversible ESPCAs*

Some non-linear reversible ESPCAs also generate interesting patterns such as fractal-like ones and disk-like ones. Below we show three examples.

ESPCA-0dca8f   It is a non-conservative reversible ESPCA (Fig. 5.91). It generates fractal-like patterns as shown in Fig. 5.92.

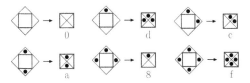

Fig. 5.91   Local function of ESPCA-0dca8f.

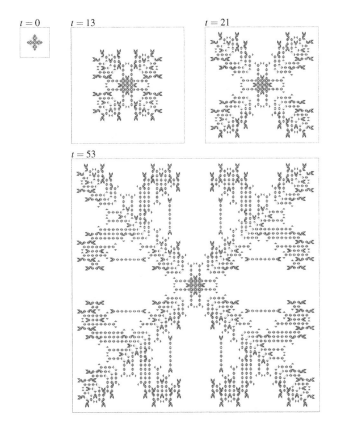

Fig. 5.92   Fractal-like patterns generated by the reversible ESPCA-0dca8f.

ESPCA-0925bf   It is a non-conservative reversible ESPCA (Fig. 5.93).
Figure 5.94 shows the patterns generated by it from the initial pattern given
in Fig. 5.88 ($t = 0$).

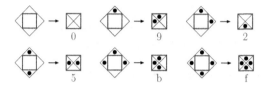

Fig. 5.93   Local function of ESPCA-0925bf.

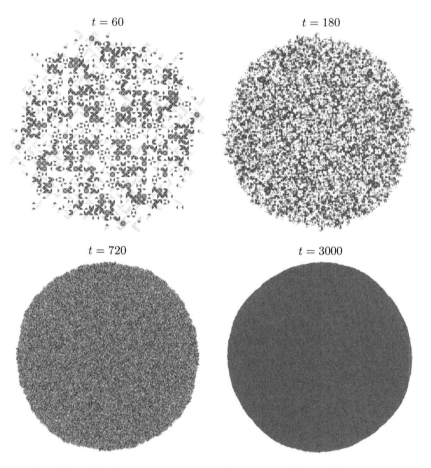

Fig. 5.94   Disk-like patterns generated by the reversible ESPCA-0925bf.

ESPCA-0925bf generates disk-like patterns that are very close to true disks. In the pattern at $t = 720$, the deviation of the distance between the center cell and each perimeter cell from the average is within $\pm 3.1\%$. In the pattern at $t = 3000$, the deviation is within $\pm 1.5\%$. It has not been clarified why this ESPCA generates these disk-like patterns.

ESPCA-0925bf is not T-symmetric under the four kinds of involutions proposed in Sec. 3.2 [39]. It is not known whether there is an involution under which ESPCA-0925bf is T-symmetric. If we go backward the evolution process of Fig. 5.94, square patterns appear and they grow indefinitely (Fig. 5.95).

$$t = -30$$

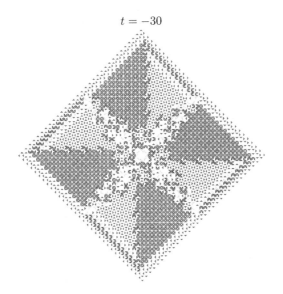

Fig. 5.95　In the negative time direction, square patterns appear in ESPCA-0925bf.

ESPCA-01eacf　It is a non-conservative reversible ESPCA (Fig. 5.96). Figure 5.97 shows the patterns generated by it.

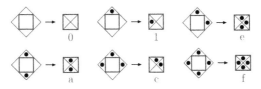

Fig. 5.96　Local function of ESPCA-01eacf.

As in the case of ESPCA-0925bf, ESPCA-01eacf generates disk-like patterns. In the pattern at $t = 720$, the deviation of the distance between the center cell and each perimeter cell from the average is within $\pm 9.5\%$. In the pattern at $t = 3000$, the deviation is within $\pm 3.7\%$. Again, it has not been clarified why this ESPCA generates these disk-like patterns.

From the local function f_{01eacf} (Fig. 5.96) of ESPCA-01eacf, we can verify $f^{\mathrm{r}}_{01eacf} = f^{-1}_{01eacf}$. By Theorem 3.2, ESPCA-01eacf is T-symmetric under the involution $H^{\mathrm{rev}} \circ H^{\mathrm{refl}}$. Therefore, it generates disk-like patterns also to the negative time direction.

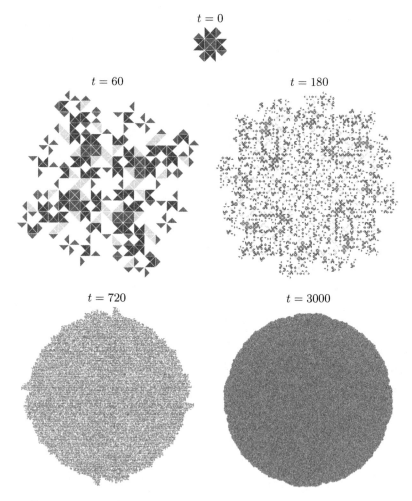

Fig. 5.97   Disk-like patterns generated by the reversible ESPCA-01eacf.

5.6.3 Growing patterns in linear ETPCAs

Linear TPCAs and ETPCAs are defined similarly as in the case of linear SPCAs and ESPCAs (see Definitions 5.1 and 5.2).

Definition 5.3. Let $\mathbb{Z}_m = \{0, 1, \ldots, m-1\}$. Let P be a TPCA whose state set is $\mathbb{Z}_m^3$ (*i.e.*, each part has the state set $\mathbb{Z}_m$), and $f : \mathbb{Z}_m^3 \to \mathbb{Z}_m^3$ be its local function. P is called a *linear TPCA* if the following holds.

$$\exists\, a_1, a_2, a_3 \in \mathbb{Z}_m :$$
$$\forall\, l, b, r, l', b', r' \in \mathbb{Z}_m :$$
$$\text{if } f(l, b, r) = (l', b', r') \text{ then}$$
$$l' = a_1 l + a_2 b + a_3 r \pmod{m}$$
$$b' = a_1 b + a_2 r + a_3 l \pmod{m}$$
$$r' = a_1 r + a_2 l + a_3 b \pmod{m}$$

Definition 5.4. A linear TPCA P is called a *linear ETPCA*, if its state set is $\{0, 1\}^3$. In this case, the addition is $\oplus$, the mod 2 addition.

Each linear ETPCA is defined by $(a_1, a_2, a_3) \in \{0, 1\}^3$, and thus there are eight linear ETPCAs. They are shown in Table 5.2. Local functions of the eight linear ETPCAs are shown in Fig. 5.98. The following two pairs of linear ETPCAs are dual under reflection: (0167, 0437) and (0330, 0660). Hence, They are essentially the same. There are three reversible ESPCAs in the eight linear ones, which are all conservative.

Table 5.2   Eight linear ETPCAs

(a_1, a_2, a_3)	ETPCA-$wxyz$	Reversible	Conservative
(0,0,0)	0000		
(0,0,1)	0167	✓	✓
(0,1,0)	0437	✓	✓
(0,1,1)	0550		
(1,0,0)	0257	✓	✓
(1,0,1)	0330		
(1,1,0)	0660		
(1,1,1)	0707		

In the following, we consider three linear irreversible ESPCAs 0707, 0330 and 0550. ETPCA-0707 show self-replication of patterns similar to the Fredkin CA (see Example 1.1), ETPCA-0330 generates snowflake-like patterns, and ETPCA-0550 generates fractal-like ones.

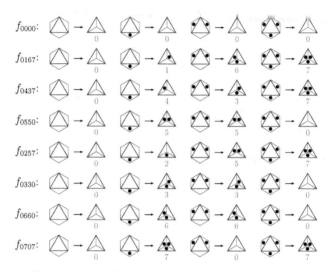

Fig. 5.98   Local functions of the eight linear ETPCAs.

ETPCA-0707   It is a linear irreversible ETPCA. Similar to the case of ESPCA-0f00f0, any initial pattern consisting of the state (1,1,1) produces its copies indefinitely (Fig. 5.99).

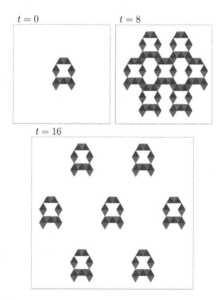

Fig. 5.99   Self-replication of a pattern in the linear ETPCA-0707.

ETPCA-0330   It is also a linear irreversible ETPCA. Starting from a single cell of the state (1,1,1), it generates snowflake-like patterns as shown in Fig. 5.100

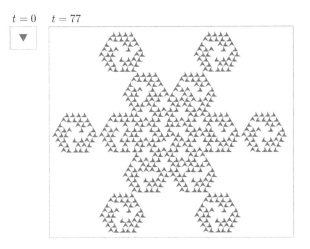

Fig. 5.100   Snowflake-like pattern generated by the linear ETPCA-0330.

If an initial pattern consists only of the states (1,1,0), (0,1,1) and (1,0,1), then its copies will appear, but other patterns are also contained (Fig. 5.101).

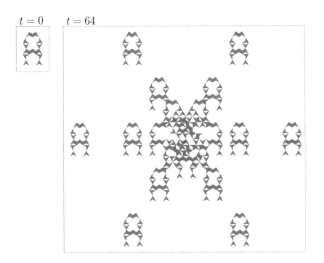

Fig. 5.101   Self-replication of a pattern in the linear ETPCA-0330.

ETPCA-0550   It is also a linear irreversible ETPCA. It generates fractal-like patterns as shown in Fig. 5.102.

$t = 0$ $t = 126$

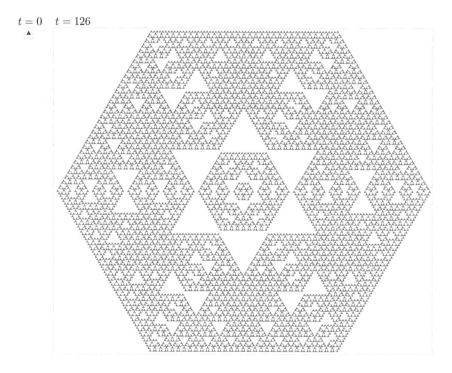

Fig. 5.102   Fractal-like pattern generated by the linear ETPCA-0550.

5.7   Remarks and Notes

We saw a variety of fantastic phenomena in reversible ESPCAs and ETP-CAs. Most patterns shown in this chapter can be seen on Golly using the files given in [21]. Since there are many kinds of ESPCAs and ETPCAs, readers may find still more interesting phenomena in them.

Similar to the Game of Life (GoL), there are many periodic patterns and space-moving patterns (or gliders) in various reversible PCAs. Interactions of such patterns often show unexpected evolution processes. Gliders are particularly important, since they can be used as information carriers. They play important roles when we compose universal computers in CAs. Gliders exist even in 1-dimensional CAs [47, 61]. Cook [47] showed Turing universality of the 1-dimensional *elementary cellular automaton* (ECA) [38] of rule 110 using several gliders and their interactions. In reversible ESPCAs and ETPCAs, gliders are used to compose RLEMs, from which reversible Turing machines are constructed, as shown in the following chapters.

Glider guns also exist in several reversible ESPCAs and ETPCAs. We observe that some of these guns generate gliders both in positive and negative time directions.

On the other hand, because of reversibility, some features of their evolving processes are very different from those in GoL, an irreversible CA. In particular, T-symmetry in reversible PCAs makes it easy to obtain a backward evolution process of a given process. We can also see that in reversible and conservative PCAs, the period of a periodic or space-moving pattern can be quite long even when the size of the pattern is small.

In Sec. 5.6, pattern generation capability in non-conservative ESP-CAs and ETPCAs is investigated. It is known that traditional types of CAs often generate fascinating patterns (see *e.g.*, [62]). Our reversible PCAs also generate interesting fractal-like patterns, disk-like ones, and others, despite the constraint of reversibility. We also studied linear PCAs, which are not necessarily reversible. Several linear PCAs show self-replication of a given initial pattern.

In the ESPCAs and ETPCAs studied in this chapter, ESPCA-02c5yf ($y \in \{b, d\}$) was first investigated in [25], ESPCA-01caef was in [26], ESPCA-01c5ef was in [28], ETPCA-0157 was in [30], ETPCA-0137 was in [63], and ETPCA-0347 was in [33]. Relation between GoL and ETPCA-0347 was discussed in [64].

5.8 Exercises

5.8.1 *Paper-and-pencil exercises*

Exercise 5.1.* Consider ESPCA-0945df (Fig. 5.21) and the periodic pattern of period 60 shown in Fig. 5.24. Write its evolving process for $0 \le t \le 14$, and observe how the population grows.

Exercise 5.2.** Consider ESPCA-05f050 (Fig. 5.83), a linear irreversible ESPCA. Prove that any initial pattern consisting of (0,0,0,0), (0,1,0,1), (1,0,1,0) and (1,1,1,1) generates its copies indefinitely as in Fig. 5.85.

Note: It can be proved in a similar way as in the Fredkin's self-replicating CA (see Exercise 1.2), but its calculation is rather complex.

5.8.2   *Golly exercises*

Exercise 5.3.* Consider ESPCA-01c5ef (Fig. 5.15). Find a space-moving pattern other than the glider-12 (Fig. 5.41) in it.

Hint: Space-moving patterns in a reversible and conservative ESPCA are often generated starting from a random-like pattern (say, of size 10×10).

Exercise 5.4.** Consider ETPCA-0347. The pattern shown in Fig. 5.82 is a glider gun that generates six glider-6's every 24 steps. Design a glider gun that generates n glider-6's every 24 steps, where $n > 6$.

Chapter 6

Making Reversible Turing Machines in Reversible ESPCAs

We study how a reversible Turing machine (RTM), a model of a reversible computer, can be designed and realized in the cellular spaces of reversible ESPCAs. Here, we consider four kinds of ESPCAs 01c5ef, 01caef, 02c5df, and 02c5bf. In each of these ESPCAs, we first search for small patterns from which interesting and useful phenomena appear. The most important pattern is one that works as a *signal*. A small glider with a short period is suited for this purpose. Since a signal should be routed from any point to any other point, we need a stable or periodic pattern(s) that controls the moving direction (and the phase) of a signal. Namely, interacting with this pattern, a signal makes a left or right turn.

Logical operation is performed in either of the two methods. The first method is applied when the following phenomenon is found: By interacting a signal with a small periodic pattern called a *position marker*, the latter is shifted by a constant distance. It works as a kind of memory, where a memory state is kept by its position. The second method is applied when the following phenomenon is found: If we interact two signals appropriately, then they appear again after the interaction, and the moving directions of them are changed. By this, a kind of a *reversible logic gate* is realized. In the following, a position marker is used in ESPCAs 01c5ef and 01caef, while a reversible logic gate is used in ESPCAs 02c5df and 02c5bf to realize a rotary element (RE). In each of these ESPCAs, an RE is implemented using only a few kinds of small patterns, and a few useful phenomena.

Since any RTM is realized by an RE, the four ESPCAs are Turing universal. Furthermore, intrinsic universality of ESPCA-02c5bf is shown (Sec. 6.4.4). If a universal RLEM is implemented in a reversible ESPCA, then its intrinsic universality is derived in a similar manner as in ESPCA-02c5bf. Thus, the above four ESPCAs are intrinsically universal.

6.1 ESPCA-01c5ef

Figure 6.1 shows the local function of ESPCA-01c5ef. It is a reversible and conservative ESPCA.

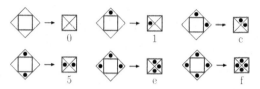

Fig. 6.1   Local function f_{01c5ef} of ESPCA-01c5ef.

6.1.1   *Useful patterns in ESPCA-01c5ef*

In ESPCA-01c5ef, there are three useful patterns called a blinker, a rotor and a glider-12. Interacting these patterns, interesting phenomena are observed. We shall see that RLEMs are constructed using the three patterns and three useful phenomena.

A *glider-12* is a space-moving pattern of period 12 (Fig. 6.2). It consists of four particles and travels one cell diagonally in 12 steps. The pattern at time p ($0 \leq p \leq 11$) (or its rotated one by a multiple of 90 degrees) is called a glider-12 of phase p. There are also other space-moving patterns in ESPCA-01c5ef (see Exercise 5.3). Among them, a glider-12 has a short period, and gives interesting and useful phenomena when interacting with other patterns. Thus, it is used as a signal when we construct RLEMs.

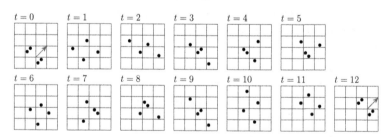

Fig. 6.2   Glider-12, a space-moving pattern of period 12 in ESPCA-01c5ef.

We use two kinds of periodic patterns. A *blinker* is a periodic pattern of period 2 shown in Fig. 6.3. A *rotor* is a periodic pattern of period 4 shown in Fig. 6.4. These patterns are used to control the moving direction of a signal. A rotor is also used as a position marker.

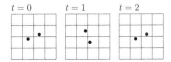

Fig. 6.3   Blinker, a periodic pattern of period 2 in ESPCA-01c5ef.

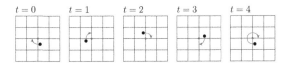

Fig. 6.4   Rotor, a periodic pattern of period 4 in ESPCA-01c5ef.

A *block* is a stable pattern shown in Fig. 6.5. When composing RLEMs and RTMs in ESPCAs-01c5ef, it will be used only for writing comments and indicating border of a module. Hence it has no functional role for processing signals. Note that in ESPCAs 02c5bf and 02c5df a block will be used to control the moving direction of a signal (see Secs. 6.3.2 and 6.4.1).

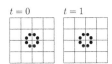

Fig. 6.5   Block, a stable pattern in ESPCA-01c5ef. It is not used for signal operations, but for indicating boundaries of modules and for writing comments.

6.1.2   *Useful phenomena in ESPCA-01c5ef*

Interacting a glider-12 with a blinker and a rotor, we obtain interesting phenomena that can be used to compose RLEMs [28]. It is remarkable that only three patterns and three phenomena shown below are sufficient to implement RLEMs.

The first is shown in Fig. 6.6. Colliding a glider-12 with a rotor in this way, the position of the rotor is shifted by 4 cells to the right. This phenomenon is used to implement a kind of memory, where the memory states are distinguished by the positions of a rotor. Changing the state of the memory is performed by this process. The rotor used for this purpose is called a *position marker*. Shifting the position marker from the right to the left is done by colliding a glider-12 from the opposite side.

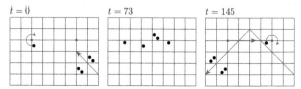

Fig. 6.6   Shifting a rotor by a glider-12 in ESPCA-01c5ef. This mechanism is called a shifting gadget. The rotor can play the role of a position marker for realizing a memory.

The second is shown in Fig. 6.7. Colliding a glider-12 with a rotor in this way, the glider-12 makes a left-turn, and after this process the rotor rotates at the same position as if it is not affected at all. This phenomenon is used to control the moving direction of a glider-12. It is also used to know the memory state (it will be explained later).

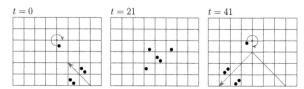

Fig. 6.7   Left-turn of a glider-12 by a rotor in ESPCA-01c5ef called a left-turn gadget.

The third is shown in Fig. 6.8. Colliding a glider-12 with a blinker, the glider-12 makes a right-turn, and as a result the blinker stays at the same position as if it is not affected at all. Although a right-turn is possible by three successive left-turns shown in Fig. 6.7, this method needs a more space. When implementing a memory using the phenomenon of Fig. 6.6, we need both right-turns and left-turns, since many access paths to a position marker must be placed in a small area. It is also used to adjust the phase of a glider-12 by combining right-turns with left-turns.

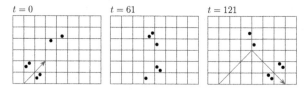

Fig. 6.8   Right-turn of a glider-12 by a blinker in ESPCA-01c5ef. It is called a right-turn gadget.

In the following, the above three phenomena are used as *gadgets*. They are called a *shifting gadget*, a *left-turn gadget*, and a *right-turn gadget*.

6.1.3   *Making RLEMs 2-3 and 2-4 in ESPCA-01c5ef*

Before composing an RE, we first show how RLEM 2-3 (Fig. 6.9) and RLEM 2-4 (Fig. 6.11) are realized in ESPCA-01c5ef. This is because they are simpler than an RE, and thus usage of a position marker and gadgets is understood more easily than the case of an RE.

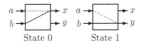

Fig. 6.9   RLEM 2-3.

Figure 6.10 shows a pattern that simulates RLEM 2-3. It consists of one position marker, five left-turn gadgets, and ten right-turn gadgets. Two small circles around the middle of the pattern show possible position of a position marker (*i.e.*, a rotor). If the position marker is in the left (right, respectively) circle, then we assume the RLEM is in the state 0 (state 1). A glider-12 is given to the input port a or b.

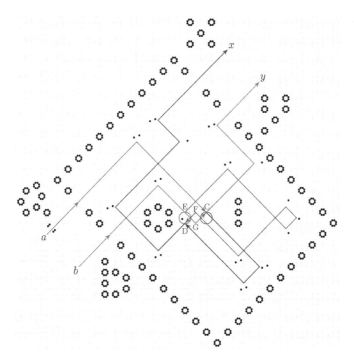

Fig. 6.10   RLEM 2-3 implemented in ESPCA-01c5ef.

We explain how the pattern simulates RLEM 2-3. In the following, $_>$P, P$_<$, and $_>$P$_<$ mean that P is a *merge point* where two signal paths merge, a *branch point* where a signal path branches into two, and a *merge-and-branch point* where both merge and branch occur, respectively.

First, consider the case where the state is 0 and an input signal (*i.e.*, glider-12) is given to a. Then the signal moves along the following path.

$$\text{Case } (0, a) : a \to C_< \to {}_>D \to x$$

The signal from a branches at C depending on the state of the RLEM. In this case, the state is 0 , and thus the signal goes straight ahead at C. Then it makes a left-turn at D. The point D acts as a merge point. Namely, this signal path merges with that of the Case $(0, b)$ at D as explained next. Finally, the signal goes out from the output port x.

Second, consider the case where the state is 0 and an input signal is given to b. Then the signal moves along the following path.

$$\text{Case } (0, b) : b \to E_< \to F \to {}_>D \to x$$

The signal coming from b branches at E depending on the state of the RLEM. Since the state is now 0, the signal makes a left-turn at E. At F the signal shifts the position marker to the right. It means the state changes to 1. The signal itself makes a left-turn at F. At D, it goes straight ahead, and thus the signal path merges with that of the case $(0, a)$. Finally the signal goes out from the output port x.

Third, consider the case where the state is 1 and an input signal is given to a. The signal moves along the following path.

$$\text{Case } (1, a) : a \to {}_>C_< \to y$$

At the branch point C, the signal makes a left-turn, since the state is 1. Thus, it does not change the state. The point C also acts as a merge point, since the signal path merges with that of the case $(1,b)$. Finally, the signal goes out from the output port y.

Fourth, consider the case where the state is 1 and an input signal is given to b. Then the signal moves along the following path.

$$\text{Case } (1, b) : b \to E_< \to G \to {}_>C \to y$$

At the branch point E, the signal goes straight ahead, since the state is 1. Then it shifts the position marker to the left. By this, the state changes to 0. The signal itself makes a left turn at G. It finally goes out from y.

By above, we can see that Fig. 6.10 correctly simulates RLEM 2-3.

Next, we compose a pattern of ESPCA-01c5ef that simulates RLEM 2-4 (Fig. 6.11). It is easily obtained from that of RLEM 2-3 using T-symmetry (Chap. 3) of ESPCA-01c5ef. Let $\delta_{2\text{-}3}$ and $\delta_{2\text{-}4}$ be the move functions of RLEM 2-3 and 2-4. They are the following mappings:

$$\delta_{2\text{-}3}: \quad (0, a) \mapsto (0, x),\ (0, b) \mapsto (1, x),\ (1, a) \mapsto (1, y),\ (1, b) \mapsto (0, y)$$
$$\delta_{2\text{-}4}: \quad (0, x) \mapsto (0, a),\ (0, y) \mapsto (1, b),\ (1, x) \mapsto (0, b),\ (1, y) \mapsto (1, a)$$

We can see $\delta_{2\text{-}4} = \delta_{2\text{-}3}^{-1}$. It means RLEM 2-4 performs the inverse operations of RLEM 2-3. By Theorem 3.2, ESPCA-01c5ef is T-symmetric under the involution $H^{\mathrm{rev}} \circ H^{\mathrm{refl}}$, since $\mathrm{inv}(01c5ef) = \mathrm{r}(01c5ef)$ holds (see Fig. 3.3). Thus, the pattern that simulates RLEM 2-4 is obtained by applying $H^{\mathrm{rev}} \circ H^{\mathrm{refl}}$ to the pattern for RLEM 2-3 (Fig. 6.10). It is shown in Fig. 6.12.

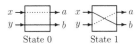

Fig. 6.11   RLEM 2-4.

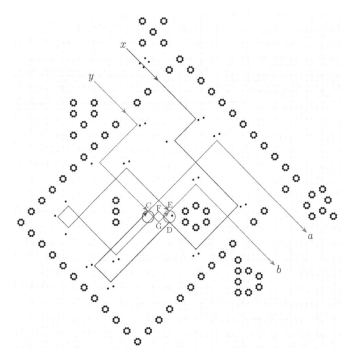

Fig. 6.12   RLEM 2-4 implemented in ESPCA-01c5ef. It is obtained by applying the involution $H^{\mathrm{rev}} \circ H^{\mathrm{refl}}$ to the main part (*i.e.*, inside the border) of the pattern in Fig. 6.10.

We can make a pattern that simulates RLEM 3-10 from the patterns of RLEMs 2-3 and 2-4 as shown in Fig. 6.13, which realizes the circuit given in Fig. 4.18. Therefore, it is possible to compose a pattern that simulates an RE out of eight copies of the pattern of RLEM 3-10 as shown in Fig. 4.17. In Sec. 6.1.4, however, we give a direct composing method of an RE in ESPCA-01c5ef using two position markers to keep the state of the RE.

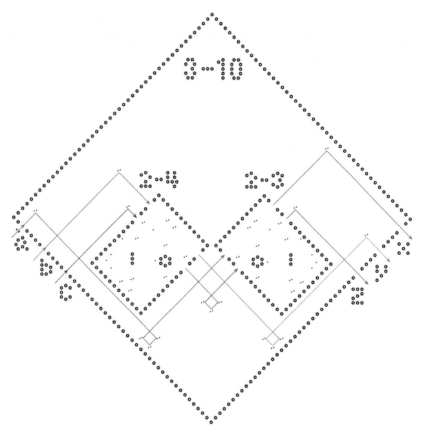

Fig. 6.13   RLEM 3-10 implemented in ESPCA-01c5ef. It is composed of the patterns of RLEMs 2-3 and 2-4.

6.1.4 *Making RE in ESPCA-01c5ef*

We now implement an RE (see Figs. 4.7 and 4.8) in ESPCA-01c5ef. A pattern that simulates an RE is shown in Fig. 6.14. It has many left-turn and right-turn gadgets. Since an RE has four input ports, we require four signal paths to know the state of an RE. Here, two position markers are used to keep the state of the RE for providing sufficient number of access paths to the markers. The four small circles labeled by V and H near the center of the pattern indicate possible positions of the markers. This figure shows the case of state V, since both of the two position markers are in the circles labeled by V. If they are in the circles labeled by H, we assume it is in the state H.

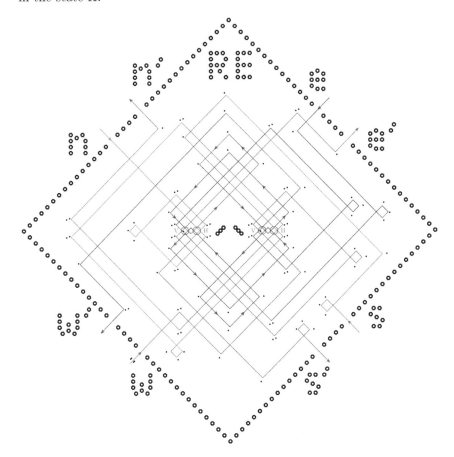

Fig. 6.14   Rotary element (RE) implemented in ESPCA-01c5ef [28].

We explain how the RE pattern works using Fig. 6.15, which is an enlarged figure of the center part of Fig. 6.14.

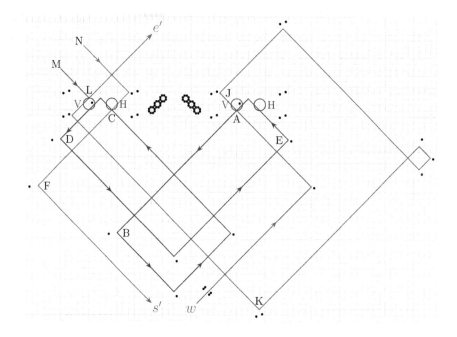

Fig. 6.15   Trajectories of a signal in the RE implemented in ESPCA-01c5ef (Fig. 6.14) in the case where the input is w. In this figure, position markers are placed in the circles labeled by V. The point A is a branch point, while C and L are merge points.

First, consider the case where the state is V and the input is w. The input signal moves along the following path:

Case (V, w) : $w \to A_< \to B \to C \to D \to E \to A \to B \to {}_>C \to F \to s'$

The signal first moves from w to A. Since the right position marker is in the circle V, the signal makes a left-turn at A. Hence, A is a branch point. The signal moves to B, and then C. At C, the signal shifts the left position marker from V to H. After that the signal goes to D, and then E. The signal again shifts the right position marker from V to H. By this, the state changes from V to H. Then it travels along the path A $\to$ B $\to {}_>$C. At C it makes a left-turn, since the left position marker is at H at this moment. The path merges with the path of the case (V, n) below at C. Finally, the signal goes out from s' via F.

Second, consider the case where the state is H and the input is w. The input signal moves along the following path:

Case (H, w) : $w \to A_< \to J \to K \to {}_>L \to e'$

The signal goes straight ahead at A, since the right position marker is at the position H. It then goes along the path J → K → L. Since the left position marker is also at the position H, the signal goes straight ahead at L, and thus goes out from e'.

In both cases of (V,w) and (H,w) the input signal first visits A. At this point the signal branches, and takes different paths depending on the state V or H. Likewise, for each of the inputs n, e and s, there is a branching point. Thus, four input lines branch into eight different paths in total.

The eight different paths must be reversibly merged into four to create four output lines n', e', s' and w'. It is explained below.

Consider the case where the state is H and the input is s. As in the case of (V,w), the signal from s first changes the state of the RE from H to V (explanation of this process is omitted here). Then it appears at the point M in Fig. 6.15. After that the signal moves along the following path:

$$\text{Case (H, } s) : \quad M \to {}_>L \to e'$$

At L the signal makes a left-turn, since the state is V at this moment. Hence, the signal goes out from e'. In this way, this signal path merges with the path of the case (H,w) at L.

Consider the case (V, n). As in the case of (H, w), after knowing the state is V, the signal from n appears at the point N. Then the signal moves as follows:

$$\text{Case (V, } n) : \quad N \to {}_>C \to F \to s'$$

At the point C, the signal goes straight ahead, since it knows the state is V. Then, the signal goes out from the output line s'. Hence, C is a merge point for the output s'.

We can verify that all other cases works correctly in a similar manner as above. Thus, we can conclude that Fig. 6.14 simulates an RE.

6.1.5 *Making RTM T_{parity} by RE in ESPCA-01c5ef*

Since an RE is implemented as shown above, it is easy to compose an RTM if we use the method shown in Sec. 4.3.5.1. Putting copies of the RE pattern at the positions corresponding to the REs in Fig. 4.28 and connecting them appropriately, we have a configuration that simulates T_{parity} (Fig. 6.16). Full computing processes of T_{parity} in ESPCA-01c5ef can be seen on Golly [1] using the simulator given in [21]. Since we can construct any RTM by this method, ESPCA-01c5ef is Turing universal.

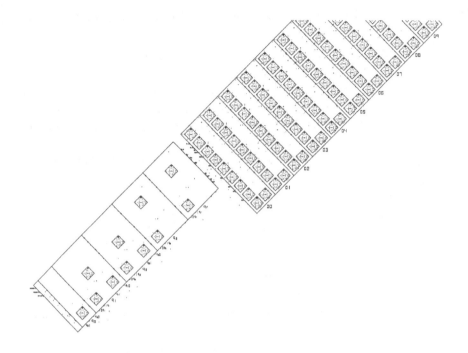

Fig. 6.16   RTM T_{parity} in ESPCA-01c5ef simulated on Golly [21]. It realizes the circuit of Fig. 4.28.

6.2   ESPCA-01caef

Figure 6.17 shows the local function of ESPCA-01caef. It is a reversible and conservative ESPCA.

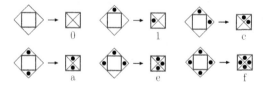

Fig. 6.17   Local function f_{01caef} of ESPCA-01caef.

6.2.1   *Useful patterns in ESPCA-01caef*

Since the local function of ESPCA-01caef (Fig. 6.17) differs from that of ESPCA-01c5ef (Fig. 6.1) only in the fourth local transition rule, some patterns, in particular a glider-12 and a blinker, evolve exactly in the same way as in ESPCA-01c5ef. However, their interactions are generally different.

We shall see that, in ESPCA-01caef, RTMs can be constructed by using only a *glider-12* (Fig. 6.18) and a *blinker* (Fig. 6.19). Note that a block (Fig. 6.5), a stable pattern, is also used to indicate borders of modules and to write comments.

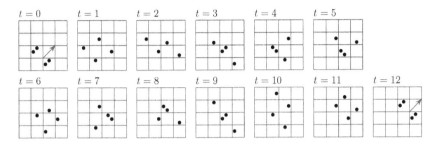

Fig. 6.18   Glider-12, a space-moving pattern, in ESPCA-01caef. It evolves in the same way as in ESPCA-01c5ef (Fig. 6.2).

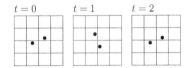

Fig. 6.19   Blinker, a periodic pattern of period 2 in ESPCA-01caef. It also evolves in the same way as in ESPCA-01c5ef.

In ESPCA-01caef, there are many kinds of space-moving patterns (Fig. 5.42). The glider-12 has the shortest period among the ones so far found. We use a glider-12 as a signal, since there are some useful phenomena related to it. Actually, it is very difficult to find such phenomena, if a space-moving pattern has a long period.

6.2.2   *Useful phenomena in ESPCA-01caef*

Using only two kinds of patterns, a glider-12 and a blinker, we can find three useful phenomena from which an RE is composed [26].

The first one is shown in Fig. 6.20. Colliding a glider-12 with a blinker in this way, the latter is shifted by six cells, and the glider-12 makes a right-turn. Thus a blinker can be used as a position marker. Note that this process takes more than 2,000 steps. In fact, its evolution process is very complex.

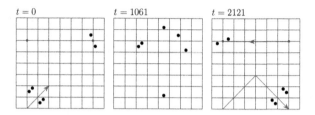

Fig. 6.20   Shifting a blinker by a glider-12 in ESPCA-01caef by six cells. It is called a shifting gadget.

Colliding a glider-12 with a blinker as in Fig. 6.21, the glider-12 makes a right-turn. Thus, a blinker can also be used to control the moving direction of a glider-12. A left-turn is performed by successive three right-turns.

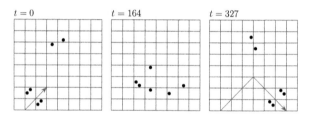

Fig. 6.21   Right-turn of a glider-12 by colliding with a blinker in ESPCA-01caef. It is called a right-turn gadget.

Colliding a glider-12 with a blinker as in Fig. 6.22, the glider-12 makes a U-turn. In the following, this gadget is used to test whether a position marker exists or not at a specified position.

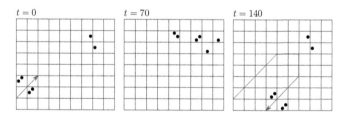

Fig. 6.22   U-turn of a glider-12 by colliding with a blinker in ESPCA-01caef. It is called a U-turn gadget.

6.2.3 *Making RLEM 2-2 in ESPCA-01caef*

As in the case of ESPCA-01c5ef (Sec. 6.1.3), before composing an RE in ESPCA-01caef, we first implement RLEM 2-2 (Fig. 6.23), since it is easier to understand. Note that RLEM 2-2 is Turing universal (Theorem 4.12), but not intrinsically universal (Theorem 4.10).

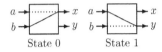

Fig. 6.23   RLEM 2-2.

Figure 6.24 shows the pattern that simulates RLEM 2-2. It uses only shifting gadgets and right-turn gadgets. The two circles around the middle of the pattern show possible positions of a position marker. If a position marker is in the left circle (right circle, respectively), then we assume the RLEM 2-2 is in the state 0 (state 1).

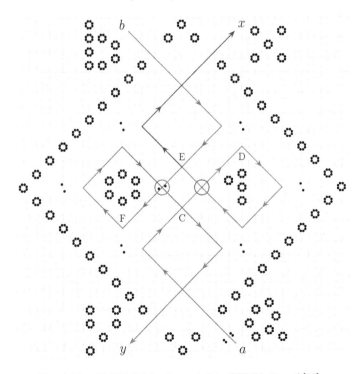

Fig. 6.24   RLEM 2-2 implemented in ESPCA-01caef [26].

We explain how the pattern simulates RLEM 2-2. We use the notations $_>$P, P$_<$, and $_>$P$_<$ to indicate that the point P is a merge point, a branch point, and a merge-and-branch point, respectively, as in Sec. 6.1.3.

When the state is 0 and the input signal is given to a, the signal moves along the following path.

$$\text{Case } (0, a) : \ a \to C_< \to D \to {}_>E \to x$$

Since the state is 0, the signal moves straight ahead at C, which is a branch point. Then it goes to D and then E. The point E is a merge point with the path of the case of $(0, b)$ shown next. Finally, it goes out from the output port x.

When the state is 0 and the input signal is given to b, the signal travels along the following path.

$$\text{Case } (0, b) : \ b \to {}_>E_< \to x$$

Since the state is 0, the signal shifts the position marker to the right at E. By this, the state of the RLEM 2-2 changes from 0 to 1. The signal itself makes a right-turn at E, and finally goes out from x. Thus, the point E works as merge-and-branch point.

When the state is 1 and the input signal is given to b, the signal moves along the following path.

$$\text{Case } (1, b) : \ b \to E_< \to F \to {}_>C \to y$$

It is symmetric with the case $(0, a)$.

When the state is 1 and the input signal is given to a, the signal travels along the following path, and the state changes from 1 to 0.

$$\text{Case } (1, a) : \ a \to {}_>C_< \to y$$

It is symmetric with the case $(0, b)$.

By above, we can see that the pattern of Fig. 6.24 correctly simulates RLEM 2-2.

6.2.4   *Making RE in ESPCA-01caef*

Figure 6.25 shows a pattern that simulates an RE. It consists of many right-turn and U-turn gadgets. In addition, four position markers are used to keep the state of the RE. The four pairs of small circles labeled by V and H indicate possible positions of the markers. If all the four markers are in the circles labeled by V (H, respectively), then the state of the RE

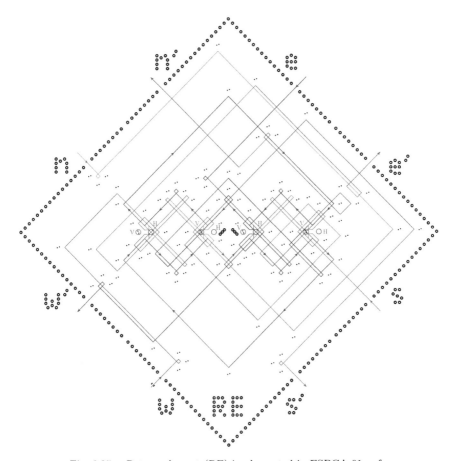

Fig. 6.25   Rotary element (RE) implemented in ESPCA-01caef.

is V (H). Therefore, when changing the state of the RE, all the position markers must be shifted to the same direction.

We explain how the pattern simulates an RE using Fig. 6.26. It contains only a part of Fig. 6.25 that is related to the case of the input is s.

First, consider the case where the state is H and the input is s. The input signal moves along the following path:

Case (H, s) :
$$s \rightarrow A_< \rightarrow B \rightarrow E \rightarrow C \rightarrow D \rightarrow E \rightarrow C \rightarrow D \rightarrow F \rightarrow {}_>G \rightarrow e'$$

The signal first moves from s to A. Since the state is H, the signal goes straight ahead at A. Hence, A is a branch point. Then the signal moves along the path $B \rightarrow E \rightarrow C \rightarrow D$. When going from D to E, the signal shifts

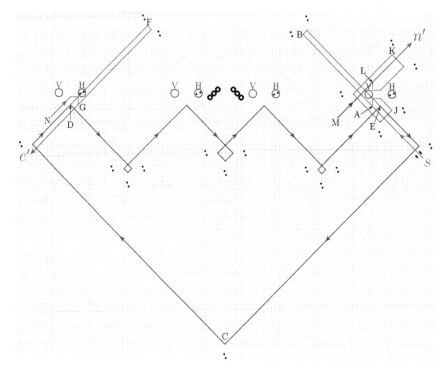

Fig. 6.26   Trajectories of a signal in the RE implemented in ESPCA-01caef in the case where the input is s. The point A is a branch point, while G and L are merge points.

all the position markers to the left. By this, the state of the RE changes to V. Then the signal moves along the path $C \rightarrow D \rightarrow F \rightarrow G$. Since the state is V at this moment, the signal moves straight at G, and goes out from e'.

Second, consider the case where the state is V and the input is s. The input signal moves along the following path:

$$\text{Case } (V, s) : \ s \rightarrow A_< \rightarrow J \rightarrow K \rightarrow {}_>L \rightarrow n'$$

The signal makes a U-turn at A, since the state is V. It then goes along the path $J \rightarrow K \rightarrow L$. Since the state is still V, the signal makes a U-turn again at L, and thus goes out from n'.

Third, consider the case (V, e). As in the case of (H,s), the signal from e first changes the state of the RE from V to H (explanation of this process is omitted here). Then it appears at the point M in Fig. 6.26. After that the signal moves along the following path:

$$\text{Case } (V, e) : \ M \rightarrow {}_>L \rightarrow n'$$

At L the signal goes straight ahead, since the state is H at this moment. Hence, the signal goes out from n'. In this way, this signal path merges with the path of the case (V, s) at L.

Fourth, consider the case (H, w). As in the case (V, s), after knowing the state is V, the signal from w appears at N. Then it moves as follows:

$$\text{Case (H, } w) : \ N \to {}_{>}G \to e'$$

At G, the signal makes a U-turn, since the state is H. Then, the signal goes out from the output port e'. Hence, G is a merge point for the output s'.

We can verify that all other cases works correctly in a similar manner.

6.2.5 *Making RTM T_{square} by RE in ESPCA-01caef*

Using the method given in Sec. 4.3.5.1 we can embed the RTM T_{square} in Example 4.3 in the cellular space of ESPCA-01caef. Figure 6.27 shows its configuration. Full computing processes of T_{square} in ESPCA-01c5ef can be seen on Golly [1] using the simulator given in [21]. Since any RTM can be constructed in this way, ESPCA-01caef is Turing universal.

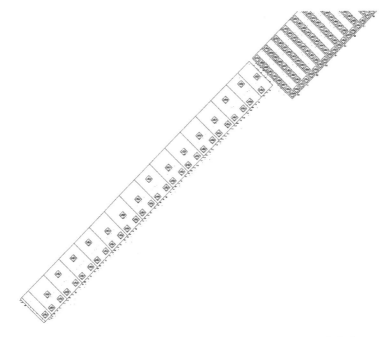

Fig. 6.27   RTM T_{square} in ESPCA-01caef simulated on Golly [21].

6.3 ESPCA-02c5df

Figure 6.28 shows the local function of ESPCA-02c5df. It is a reversible and conservative ESPCA. Different from the cases of ESPCAs 01c5ef and 01caef, no pattern that works as a position marker has been found in ESPCAs 02c5df and 02c5bf. On the other hand, there is a phenomenon that realizes a kind of a reversible logic gate called an interaction gate (I-gate). Here, an RE is constructed out of an I-gate and its inverse.

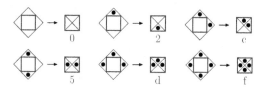

Fig. 6.28 Local function f_{02c5df} of ESPCA-02c5df.

As noted in Sec. 1.1.2.2, A block cellular automaton by Margolus [20] has the so-called Margolus neighborhood. There, all the cells are grouped into blocks of size 2×2. Figure 6.29 shows a set of block rules for a particular block CA [20]. They are applied to all the blocks in the cellular space simultaneously. At $t = 0$ an original block grouping is used, at $t = 1$ the grouping shifted by (1,1) is used, at $t = 2$ the original grouping is used again, and so on, alternately. He showed that a Fredkin gate, a universal reversible logic gate, is realized in this CA.

We can see that local functions in Figs. 6.28 and 6.29 are isomorphic. In fact, any evolution process in ESPCA-02c5df is simulated in the Margolus CA. Therefore, the following construction of an RE and RTMs are also possible in the Margolus CA.

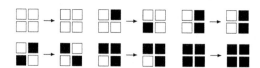

Fig. 6.29 Local function for the Margolus CA consisting of block rules [20].

6.3.1 *Useful patterns in ESPCA-02c5df*

There are two useful patterns in ESPCA-02c5df. They are a glider-1w for a signal, and a block for controlling it.

Figure 6.30 shows a *glider-1w* that is a space-moving pattern of period 1. It shifts its position by one cell in one step. Though a one-particle pattern (*i.e.*, glider-1 in Fig. 5.39) also acts so, the glider-1w consists of two particles, since a kind of logical operation is performed by this as seen below.

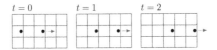

Fig. 6.30   *Glider-1w*, a space-moving pattern of period 1 in ESPCA-02c5df.

Figure 6.31 shows a *block*, a stable pattern. It is the same pattern given in ESPCA-01c5ef (Fig. 6.5) and also used in ESPCA-01caef. However, here, it is used for changing the moving direction of a glider-1w, besides indicating boundaries of modules and writing comments as in the cases of ESPCA-01c5ef and ESPCA-01caef.

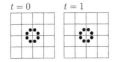

Fig. 6.31   *Block*, a stable pattern in ESPCA-02c5df.

6.3.2   Useful phenomena in ESPCA-02c5df

There are two useful phenomena in ESPCA-02c5df. They are a left/right-turn of a signal by blocks, and an interaction of two signals [25].

Figure 6.32 shows that a left-turn is possible by colliding a signal with two blocks. A right-turn is realized by the mirror image of this figure. Therefore, it is called a *left/right-turn gadget*.

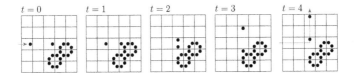

Fig. 6.32   Left-turn of a glider-1w by blocks in ESPCA-02c5df. Since a right-turn is possible similarly, it is called a left/right-turn gadget.

Figure 6.33 shows a collision of two signals. By this, a kind of logical operation is performed. This process mimics a collision of two elastic balls in the billiard-ball model (BBM) shown in Fig. 6.34 proposed by Fredkin and Toffoli [52].

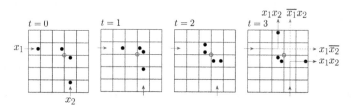

Fig. 6.33   Collision of two glider-1w's in ESPCA-02c5df.

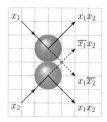

Fig. 6.34   Collision of two balls in the billiard ball model (BBM) [52].

The processes in Figs. 6.33 and 6.34 realize a reversible logic gate called an *interaction gate* (I-gate) [52] shown in Fig. 6.35(a). An I-gate is a 2-input 4-output logic gate having the following injective logic function $f_I : (0,0) \mapsto (0,0,0,0)$, $(0,1) \mapsto (0,1,0,0)$, $(1,0) \mapsto (0,0,1,0)$, $(1,1) \mapsto (1,0,0,1)$.

By the inverse process of a collision of two glider-1w's (or two balls), we obtain a 4-input 2-output reversible logic gate called an *inverse interaction gate* (I^{-1}-gate). It is a logic gate having the following partial injective logic function $f_I^{-1} : (0,0,0,0) \mapsto (0,0)$, $(0,1,0,0) \mapsto (0,1)$, $(0,0,1,0) \mapsto (1,0)$, $(1,0,0,1) \mapsto (1,1)$ (Fig. 6.35(b)).

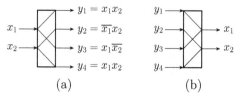

Fig. 6.35   (a) Interaction gate (I-gate), and (b) its inverse gate (I^{-1}-gate).

6.3.3 *Making RE by interaction gate in ESPCA-02c5df*

It is possible to compose an RE out of four I-gates and four I^{-1}-gates. The circuit that simulates an RE is given in Fig. 6.36 [28]. In it, many *delay elements* are also used to adjust the signal timing. They are indicated by small triangles, in which a delay time is shown by an integer.

To keep the state V or H of an RE, we use a *state signal* that moves along the circular path labeled by V or H in Fig. 6.36. If there is no input signal, the state signal circulates with the period of 4 units of time. When we give an input signal, it must be done at time $t \equiv 0 \pmod{4}$.

When two signals come to an I-gate or an I^{-1} gate, they must arrive exactly at the same time. Therefore, when implementing the circuit of Fig. 6.36 in ESPCA-02c5df, we must insert an appropriate delay in each signal line. This is indeed a cumbersome task. However, once a pattern that simulates an RE is obtained in this way, making larger functional modules out of REs becomes easier. If the period of the state signal in the RE pattern is p, then an input signal can be given to the RE at *any* time t that satisfy $t \equiv 0 \pmod{p}$ (in the patter given in Fig. 6.40, $p = 1,000$).

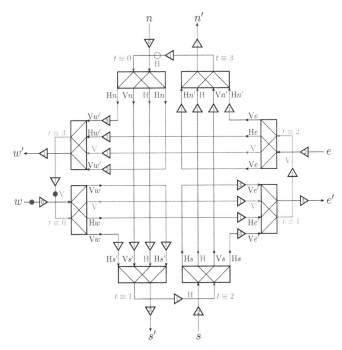

Fig. 6.36   Composing an RE using I-gates, I^{-1}-gates, and delay elements [28].

We explain how the circuit of Fig. 6.36 works.

Consider the case where the state is V and an input signal is given to w at $t \equiv 0$ (mod 4). The state signal and the input signal interact at the west I-gate. The two signals come out from the two output ports of the west I-gate labeled by Vw. They reach the two input ports Hs' of the south I^{-1}-gate. Then, one of the signal enters the cycle labeled by H. By this, the state changes from V to H correctly. The other signal goes out from the output port s' at $t + 4$.

Next, consider the case where the state is H and an input signal is given to w. Since the state signal moves along the cycle labeled by H, the input signal does not interact with it. Hence, the state does not change. The input signal comes out from the output port Hw of the west I-gate. This signal reaches the input port He' of the east I^{-1}-gate at $t + 1$, and then goes out from e' at $t + 4$.

Since other cases are similar, we can see the circuit simulates an RE.

We explain a method of implementing the circuit in ESPCA-02c5df.

Although an I-gate is realized in ESPCA-02c5df by a collision of two glider-1w's as shown in Fig. 6.33, it is convenient to make it as an *I-gate module* so that the delays between the inputs and the outputs become constant. Figure 6.37 is an I-gate module whose delay is 48 steps.

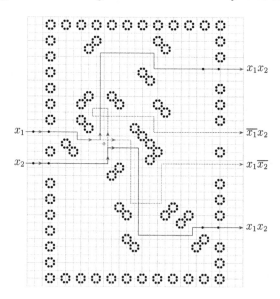

Fig. 6.37   I-gate module in ESPCA-02c5df [28].

Figure 6.38 shows the pattern of the I^{-1}-*gate module*, which is obtained by applying the involution $H^{\text{rev}} \circ H^{\text{refl}}$ to the pattern in Fig. 6.37. This is because $\text{inv}(02\text{c}5\text{df}) = \text{r}(02\text{c}5\text{df})$ holds (see Fig. 3.3), and thus ESPCA-02c5df is T-symmetric under $H^{\text{rev}} \circ H^{\text{refl}}$ by Theorem 3.2.

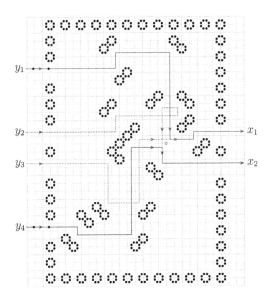

Fig. 6.38 I^{-1}-gate module in ESPCA-02c5df [28].

We also need various kinds of *delay modules*. Figure 6.39 shows an example of a delay module with an additional delay of 256 steps.

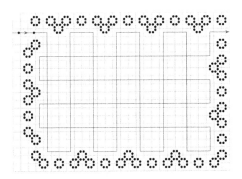

Fig. 6.39   Example of a delay module in ESPCA-02c5df [28]. It has an additional delay of 256 steps.

Assembling copies of the I-gate, the I^{-1}-gate and delay modules and connecting them as in Fig. 6.36 we have a pattern that simulates an RE in ESPCA-02c5df. Figure 6.40 shows the pattern of an RE module simulated on Golly. In this implementation, the state signal circulates with the period 1000. Therefore, an input signal must be given at $t \equiv 0 \pmod{1000}$. Note that in this figure, besides the 16 states of ESPCA-02c5bf, some additional states are used for indicating the state V or H of the RE. Otherwise, it is difficult to recognize the state.

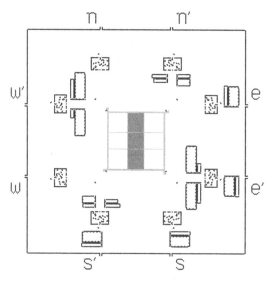

Fig. 6.40   RE realized in ESPCA-02c5df simulated on Golly. It is in the state V.

6.3.4   *Making RTM T_{prime} by RE in ESPCA-02c5df*

By the method given in Sec. 4.3.5.1 we can embed an RTM T_{prime} that accepts the unary language $\{1^n \mid n \text{ is a prime number}\}$. Figure 6.41 shows its configuration. The RTM T_{prime} is a 54-state 2-symbol RTM, but we do not describe its details here. Full computing processes of T_{prime} in ESPCA-02c5df can be seen on Golly [1] using the simulator given in [21]. In this way, we can conclude ESPCA-02c5df is Turing universal.

Fig. 6.41   RTM T_{prime} implemented in ESPCA-02cdbf simulated on Golly [21].

6.4 ESPCA-02c5bf

Figure 6.42 shows the local function of ESPCA-02c5bf. It is a reversible and conservative ESPCA.

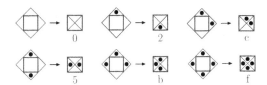

Fig. 6.42   Local function f_{02c5bf} of ESPCA-02c5bf.

6.4.1 *Useful patterns and phenomena in ESPCA-02c5bf*

The local function of ESPCA-02c5bf (Fig. 6.42) differs from that of ESPCA-02c5df (Fig. 6.28) only in the fifth local transition rule. Hence, a *glider-1w* and a *block* behave in the same way as in ESPCA-02c5df (Figs. 6.30 and 6.31). A collision of two glider-1w's is also the same (Fig. 6.33). However, left/right turn of a glider-1w is different.

A left-turn of a glider-1w is performed by colliding it with a single block as in Fig. 6.43. It is called a *left-turn gadget*. However, no right-turn is possible if we use only one block (Fig. 6.44). Thus, a right-turn is implemented by three successive left-turns.

From these observations, we can see that an RE can be implemented in a similar method as in the case of ESPCA-02c5df.

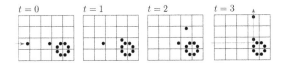

Fig. 6.43   Left-turn of a glider-1w by a block in ESPCA-02c5bf.

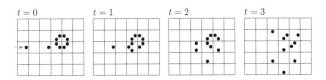

Fig. 6.44   Right-turn of a glider-1w is not possible by a block in ESPCA-02c5bf.

6.4.2   *Making RE by interaction gate in ESPCA-02c5bf*

An I-gate module is implemented as shown in Fig. 6.45. Its input-output delay is 48 steps. An I^{-1}-gate module is also obtained by applying the involution $H^{\mathrm{rev}} \circ H^{\mathrm{refl}}$ to this pattern.

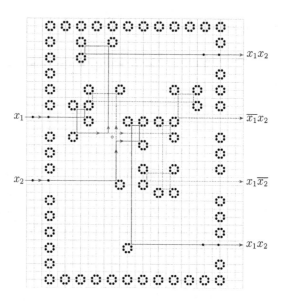

Fig. 6.45   I-gate module in ESPCA-02c5bf [28].

Figure. 6.46 shows a delay module with an additional delay of 500 steps.

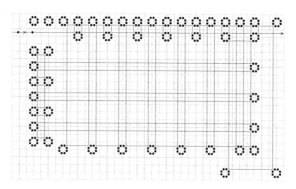

Fig. 6.46   Example of a delay module in ESPCA-02c5bf [28].

From I-gate, I^{-1}-gate, and delay modules we can construct an RE module in ESPCA-02c5bf as in the case of ESPCA-02c5df. Figure 6.47 shows its pattern. Also in this implementation, the state signal circulates with the period 1000. Therefore, an input signal must be given at $t \equiv 0$ (mod 1000).

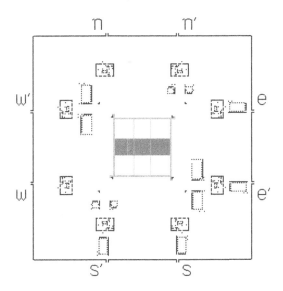

Fig. 6.47   RE realized in ESPCA-02c5bf simulated on Golly. It is in the state H.

6.4.3   *Making RTM T_{power} by RE in ESPCA-02c5bf*

Figure 6.48 is an RTM T_{power} implemented in ESPCA-02c5bf that accepts the unary language $\{1^n \mid n = 2^k \ (k = 0, 1, 2, \dots) \}$, which is given in Example 4.2. Its computing processes can be seen on Golly [1] using the simulator given in [21]. Since any RTM can be constructed by this method, ESPCA-02c5bf is Turing universal.

Fig. 6.48   RTM T_{power} implemented in ESPCA-02cdbf simulated on Golly [21].

6.4.4 *Intrinsic universality of ESPCA-02c5bf*

A *metacell* is a pattern of a CA that simulates one cell of another (or maybe the same) CA. Placing copies of the metacell to form a finite or infinite array in the original CA, evolution processes of the latter CA are simulated. Here, we give a method of composing a metacell that simulates ESPCA-01caef in the cellular space of ESPCA-02c5bf. Though ESPCA-01caef is a particular SPCA, it is possible to extend this method for creating a metacell of *any* reversible SPCA. Therefore, ESPCA-02c5bf is intrinsically universal with respect to the class of all reversible SPCAs (Definition 4.24).

The local function f_{01caef} of ESPCA-01caef (see Fig. 6.17) is the following mapping:

$$0000 \mapsto 0000, \ 0001 \mapsto 1000, \ 0010 \mapsto 0001, \ 0011 \mapsto 1100,$$
$$0100 \mapsto 0010, \ 0101 \mapsto 0101, \ 0110 \mapsto 1001, \ 0111 \mapsto 1110,$$
$$1000 \mapsto 0100, \ 1001 \mapsto 0110, \ 1010 \mapsto 1010, \ 1011 \mapsto 0111,$$
$$1100 \mapsto 0011, \ 1101 \mapsto 1011, \ 1110 \mapsto 1101, \ 1111 \mapsto 1111$$

Here, $f_{01caef}(t, r, b, l) = (t', r', b', l')$ is expressed by $trbl \mapsto t'r'b'l'$. We first compose a circuit that computes f_{01caef}. Figure 6.49 shows the circuit, where each small triangle is a *delay element* having a suitable delay time.

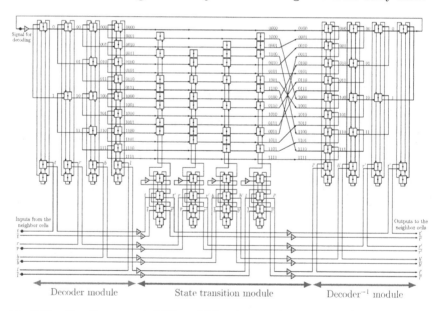

Fig. 6.49 Circuit composed of REs and delay elements that computes the local function of ESPCA-01caef. It is used to compose a metacell in ESPCA-02c5bf.

The circuit consists of three modules, *i.e.*, a *decoder module* (the left part), a *state transition module* (the center part), and an *inverse decoder module* (the right part).

When the input (t, r, b, l) is given, the decoder module determines which one of the 16 cases is occurring. Note that the inputs and outputs are represented by a dual-rail data (see Sec. 4.3.6), since ESPCA-02c5bf is conservative, and thus neither annihilation nor creation of a signal is possible. Figure 6.49 shows the case where the input (t, r, b, l) is $(1, 0, 0, 0)$. The four columns in the decoder module are RE-columns (see Sec. 4.3.4.1), and correspond to t, r, b, and l, respectively. At first, input signals t, r, b, and l are sent to the four columns. If the value of a signal is 1, then the RE-column is set to the state 1, else it remains to be in the state 0. After that, a signal for decoding, which is shown at the upper left corner, is sent to the four RE-columns. Then, depending on the states of the columns, the signal branches. Finally, it comes out from the port labeled by $trbl$. For example, if $(t, r, b, l) = (1, 0, 0, 0)$, then the signal appears at the port 1000. All the columns are (automatically) reset to 0.

If a decoded signal is given, the state transition module computes $f_{01caef}(t, r, b, l) = (t', r', b', l')$. The four columns again correspond to t, r, b, and l, respectively. The lower part of a column is an RE-column of degree 3, while the upper part is a subroutine caller (Fig. 4.26) to the RE-column having one calling port and one return port. Consider, *e.g.*, a decoded signal is given to the port 1000. Since $f_{01caef}(1, 0, 0, 0) = (0, 1, 0, 0)$, t and r should be complemented, while b and l should be kept unchanged. The decoded signal from 1000 accesses the first and the second RE-columns via the subroutine callers and set them to the state 1, but it does not access the third and the fourth. After that, a signal pair $(t, \bar{t})$ is given to the RE-column. Since the first RE-column is in the state 1, the signals t and $\bar{t}$ are exchanged, and thus we have $(t', \bar{t'}) = (\bar{t}, t)$. Likewise, $(r', \bar{r'}) = (\bar{r}, r)$. However, the third and the fourth RE-columns are in the state 0, we have $(b', \bar{b'}) = (b, \bar{b})$, and $(l', \bar{l'}) = (l, \bar{l})$. By this, the output (t', r', b', l'), which is the next state of the cell of ESPCA-01caef, is obtained correctly.

It should be noted that, to simulate a cell of ESPCA-01caef repeatedly, the signal for decoding must be recycled. Therefore, the 16 decoded channels should be reversibly merged into one. This process is performed by an inverse circuit of the decoder module. However, before doing it, the 16 decoded channels must be re-arranged as shown in the right part of the state transition module. This is because the inverse decoding will be done using the new state (t', r', b', l').

The inverse decoder module is just the mirror image of the decoder module (but, the signal direction should also be reversed). In this module, the decoded signal is first given to one of the 16 channels. Then, the 16 channels are merged into one, and the signal appears at the right end pf the circuit. But, the information from which channel the signal came is left in the bottom REs of the four RE-columns. However, it is reversibly erased by giving t', r', b', and l' signals to this circuit. By above, one cycle of simulation is completed.

Figure 6.50 shows a metacell for ESPCA-01caef implemented in ESPCA-02c5bf. Figure 6.51 is a pattern that simulates the circuit given in Fig. 6.49.

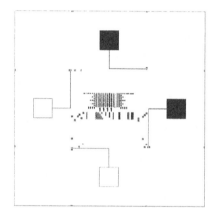

Fig. 6.50　　Metacell of size $10,000 \times 10,000$ that simulates ESPCA-01caef. It is realized in ESPCA-02c5bf on Golly. It simulates one step of ESPCA-01caef in 1,000,000 steps.

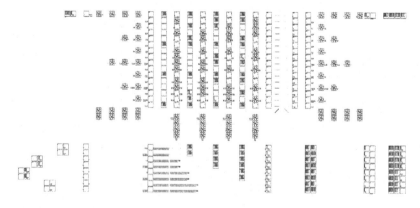

Fig. 6.51　　Center part of the metacell (Fig. 6.50) in ESPCA-02c5bf. It consists of many RE modules and delay modules, and realizes the circuit shown in Fig. 6.49.

Figure 6.52 is an evolution process of a finite array of metacells in ESPCA-02c5bf [21]. It simulates ESPCA-01caef given in Fig. 6.53. Note that, there, an array of size 3×3 is simulated, since Golly can deal with only finite configuration.

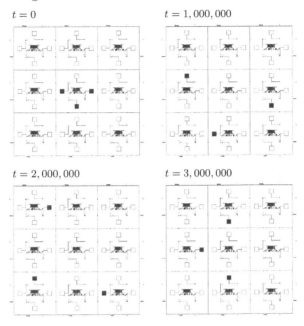

Fig. 6.52 3×3 array of metacells implemented in ESPCA-02c5bf [21]. It simulates the evolution process of ESPCA-01caef shown in Fig. 6.53.

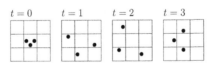

Fig. 6.53   Evolution process of a 3-dot pattern in ESPCA-01caef.

Extension of the above construction method to other reversible ESPCAs is easy. It is done by changing only the state transition module in Fig. 6.49. It is also possible to design a metacell for a reversible SPCA where the number of states of each part is more than two. For example, when the top part of a cell has four state, we use two bits of dual-rail data, say $(t_0, \overline{t_0})$ and $(t_1, \overline{t_1})$, to represent its state. If each of the four parts has four states, then 256 decoded channels are created using eight RE-columns in the decoder module. Therefore, the size of a metacell becomes quite large, but it is constructed in a similar manner as in he case of ESPCA-01caef.

6.5   Remarks and Notes

It is shown that an RE can be composed in the cellular spaces of ESPCAs 01c5ef, 01caef, 02c5df, and 02c5bf. Depending on the properties of the ESPCAs, different strategies were taken to compose it. In ESPCAs 01c5ef and 01caef, position markers were used to implement an RE, while in ESP-CAs 02c5df and 02c5bf, I-gates and I^{-1}-gates were used. However, once an RE is obtained, any RTM is systematically constructed in each ESPCA.

By above, the four reversible ESPCAs with horizontally ultimately periodic configurations are proved to be Turing universal. In addition, their dual ESPCAs under reflection are also Turing universal, since they can simulate the original ESPCAs by taking mirror images of configurations (see Lemma 3.2). Identification numbers of these ESPCAs are 04c5bf = r(01c5ef), 04cabf = r(01caef), 02c5df = r(02c5df), and 02c5ef = r(02c5bf) as it is seen in Fig. 3.3. However, so far, no other Turing universal ESPCA has been found. It is also an open problem whether a Turing universal ESPCA with *finite configurations* exists. In Chap. 8, it will be shown that we can obtain a Turing universal reversible SPCA with finite configurations, if we increase the number of states of each part by one (*i.e.*, an 81-state SPCA).

In Sec. 6.4.4, a *metacell* is constructed in ESPCA-02c5bf, which acts as a cell of ESPCA-01caef. Hence, the latter is simulated inside the former. Extending the method, it is possible to simulate *any* reversible SPCA in ESPCA-02c5bf. Hence, ESPCA-02c5bf is intrinsically universal with respect to the class of all reversible SPCAs. If a universal reversible logic gate or a universal RLEM is realizable in a reversible SPCA, then its intrinsic universality can be shown in a similar manner as in the case of ETPCA-02c5bf. Therefore, all of the above Turing universal ESPCAs are intrinsically universal. In the Game of Life CA, several kinds of metacells have been created so far (see "metacell" in [8], "unit cell" in [7], and patterns in the `HashLife/Metacell` folder of Golly [1]).

Note that a method of making RTMs in a simple reversible PCA was first given in [65] using ETPCA-0347 (see also Chap. 7). Later, it was shown in ESPCA-01caef [26], where RLEM 4-31 was used to make RTMs. A method of using REs in ESPCA-01caef was given in [27], where an RE is composed of RLEMs 2-3 and 2-4. Construction methods of an RE in ESPCAs 01c5ef, 02c5bf, and 02c5df were shown in [28].

6.6 Exercises

6.6.1 *Paper-and-pencil exercises*

Exercise 6.1.* Consider ESPCA-01c5ef (Fig. 6.54) and the process of left-turn of a glider-12 by a rotor shown in Fig. 6.7. The configuration at $t = 12$ is shown in Fig. 6.55. Write the configurations for $12 \leq t \leq 29$, and verify that the glider-12 actually makes a left-turn.

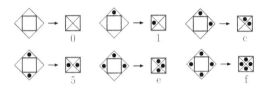

Fig. 6.54   Local function of ESPCA-01c5ef.

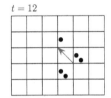

Fig. 6.55   Left-turn of a glider-12 by a rotor in ESPCA-01c5ef.

Exercise 6.2.** Consider ESPCA-01c5ef (Fig. 6.54). Design a pattern that simulates RLEM 2-2 (Fig. 6.23) in it.

Exercise 6.3.*** Consider ESPCA-01caef (Fig. 6.56). Design a pattern that simulates RLEM 2-3 (Fig. 6.9) in it.

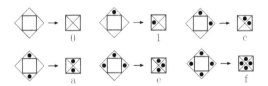

Fig. 6.56   Local function of ESPCA-01caef.

Exercise 6.4.** Design a circuit composed of I-gates, I^{-1}-gates (Fig. 6.35), and delay elements that simulates RLEM 2-2 (Fig. 6.23).
Hint: Use a similar method as the one shown in Fig. 6.36.

Exercise 6.5.*** Design a circuit composed of I-gates, I^{-1}-gates and delay elements that simulates RLEM 2-3 (Fig. 6.9).

6.6.2 *Golly exercises*

Exercise 6.6.* Consider the RTM T_{square} (Fig. 6.27) implemented in ESPCA-01caef. In the pattern file of T_{square} for Golly contained in [21], the cases where inputs are $n = 4$ and $n = 5$ are given. Change the input to other values, say $n = 9$, and simulate it on Golly.

Note: It is done by changing the state of an RE that contains a tape symbol in a tape cell module (see Fig. 4.23). Use the copy and paste functions equipped in Golly.

Exercise 6.7.** Implement the pattern of RLEM 2-2 for ESPCA-01c5ef designed in Exercise 6.2 using Golly. Then, verify that it works correctly.

Exercise 6.8.** Implement the pattern of RLEM 2-3 for ESPCA-01caef designed in Exercise 6.3 using Golly. Then, verify that it works correctly.

Exercise 6.9.*** Create patterns for ESPCA-02c5df that simulate the circuits designed in Exercises 6.4 and 6.5 using Golly.

Note: There may be a difficulty in adjusting the signal timing at each gate.

Chapter 7

Making Reversible Turing Machines in Reversible ETPCAs

We study how RTMs are constructed in reversible elementary triangular partitioned cellular automata (ETPCAs). As we have seen in Sec. 2.2, an ETPCA is described by only four local transition rules, since the number of neighboring cells is three. It is quite simple, yet some reversible ETPCAs have rich capability of computing as we shall see below. Thus, the framework of ETPCAs is useful to investigate how simple an RCA can be to be Turing universal.

Besides the standard ETPCAs, such as ETPCA-0347 and ETPCA-0157, we also consider ones whose local functions are defined *partially*. More specifically, we investigate ETPCA-034z and ETPCA-013z, where z means that the fourth local transition rule is undefined. We shall see that in these partial ETPCAs any RTM is realized by using only three local transition rules. We can interpret z as the one that stands for the two cases of $z = 0$ and $z = 7$. However, a more important fact is that these ETPCAs are Turing universal without using the fourth local transition rule.

In ETPCAs 0347, 0157 and 013z, we implement RLEM 4-31, and then construct RTMs using it by the method shown in Fig. 4.34. Since RLEM 4-31 is intrinsically universal, any RLEM, in particular an RE, can be realized by a circuit composed of it. Therefore, we can also show that these ETPCAs are intrinsically universal, *i.e.*, any reversible TPCA can be simulated in them in a similar manner as in ESPCA-02c5bf (Sec. 6.4.4).

In ETPCA-034z, we compose RLEM 2-2. Though RLEM 2-2 is not intrinsically universal (Theorem 4.10), it has been shown to be Turing universal by Cook and Palmiere (Theorem 4.12). Therefore, it is possible to construct RTMs in ETPCA-034z. However, since their method of composing RTMs out of RLEM 2-2 is quite complex, no concrete configuration of ETPCA-034z that simulates an RTM is given here.

7.1   ETPCA-0347

Figure 7.1 shows the local function of ETPCA-0347. It is a reversible but
not conservative ETPCA.

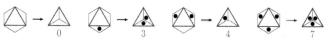

Fig. 7.1   Four local transition rules that define the local function of ETPCA-0347.

7.1.1   *Useful patterns in ETPCA-0347*

We use three patterns to construct RTMs in ETPCA-0347. The first one is
a *glider-6*, a space-moving pattern of period 6 (Fig. 7.2). It can move in six
directions, and will be used as a signal. The pattern at time p ($0 \le p \le 5$)
(or its rotated one by a multiple of 60 degrees) is called a glider-6 of phase p.
The second is a *fin*, a periodic pattern of period 6 (Fig. 7.3). It will be used
as a *position marker* similar to the cases of ESPCA-01c5ef (Sec. 6.1.2) and
ESPCA-01caef (Sec. 6.2.2). The third is a *block*, a stable pattern (Fig. 7.4).
It will be used to change the moving direction of a glider-6.

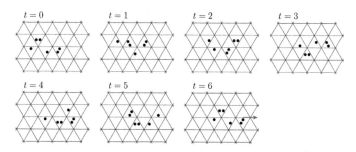

Fig. 7.2   Glider-6, a space-moving pattern of period 6 in ETPCA-0347.

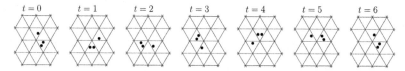

Fig. 7.3   Fin, a periodic pattern of period 6 in ETPCA-0347.

Fig. 7.4   Block, a stable pattern in ETPCA-0347.

7.1.2 Useful phenomena in ETPCA-0347

There are various phenomena that can be used to compose an RLEM. Colliding a glider-6 with two blocks as in Fig. 7.5, a right-turn by 60 degrees is performed. It is called a *60° -right-turn gadget*. Successive executions of 60°-right-turns, the moving direction of a glider is changed freely.

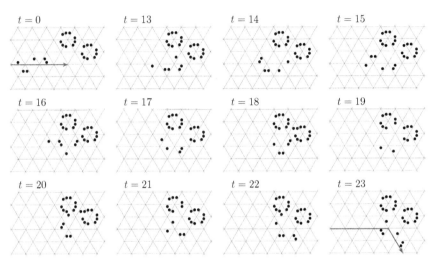

Fig. 7.5   60°-right-turn gadget of a glider-6 in ETPCA-0347 [66].

A 120°-right-turn of a glider-6 is possible using two blocks as in Fig. 7.6, which is called a *120° -right-turn gadget*. The glider-6 is first decomposed into a "body" (left) and a fin (right) ($t = 56$). The body rotates around the point indicated by a small circle, and the fin goes around the blocks. Finally, the body meets the fin, and a glider-6 is reconstructed. Then the glider-6 moves to the south-west direction ($t = 334$).

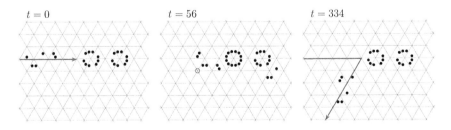

Fig. 7.6   120°-right-turn gadget of a glider-6 by two blocks in ETPCA-0347 [33].

A 120°-right-turn is also possible by using a sequence of three blocks (Fig. 5.60) or five blocks. As in the case of two blocks, a glider-6 is first decomposed into a body and a fin. The fin moves around the sequence of blocks. Finally, they meet, and a glider-6 is reconstructed. They differ only in their delay time as shown in Table 7.1.

Note that a 120°-right-turn is obtained also by two 60°-right-turns. The latter implementation is more useful, since a right-turn is performed more quickly. However, a 120°-right-turn gadget is usable as an interface between *unidirectional signal paths* and a *bidirectional signal path* as shown in Fig. 7.7. A bidirectional signal path appears, for example, when shifting a fin by a glider-6 (see Fig. 7.10(a) and (b)). In this case, such an interface is necessary.

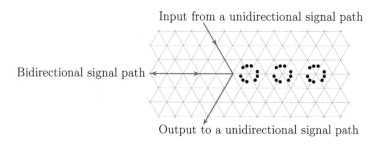

Fig. 7.7   Interface between unidirectional signal paths and a bidirectional signal path using a sequence of three blocks in ETPCA-0347 [33].

A backward-turn of a glider-6 is possible by colliding it with a block as shown in Fig. 7.8, which is called a *backward-turn gadget*. As in the case of a 120°-right-turn, a glider-6 is first split into a body (left) and a fin (right) ($t = 31$), and the fin moves around the block. Finally, the fin is attached to the body to form a glider-6 again. The created glider moves backward ($t = 97$).

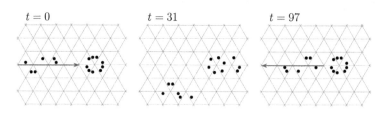

Fig. 7.8   Backward-turn gadget of a glider-6 in ETPCA-0347 [33].

A U-turn of a glider-6 is performed by a *U-turn gadget* composed of five blocks shown in Fig. 7.9. Also in this case, a glider-6 is first decomposed into a body and a fin, and finally a glider-6 is reconstructed.

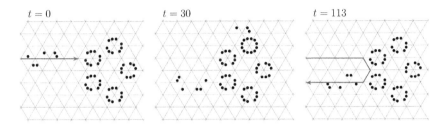

Fig. 7.9   U-turn gadget of a glider-6 in ETPCA-0347 [33].

Six kinds of turn gadgets are introduced above. Table 7.1 gives the net delay and the phase shift of these turn gadgets. The *net delay* of a turn gadget is the additional delay caused by the gadget. For example, consider the U-turn gadget (Fig. 7.9). We can see that the traveling distance of the glider-6 from $t = 0$ to $t = 113$ along the arrow line is 6 under the assumption that the side-length of a triangle is 1. Note that both of the glider-6 patterns at $t = 0$ and 113 are of phase 0, since the traveling time should be measured between two configurations having the same phase. If a glider travels the distance 6 in a vacant space, it takes $6/(1/6) = 36$ steps. Therefore, the net delay by the turn gadget is $113 - 36 = 77$. The *phase shift* s of a turn gadget is the phase shift value of a glider-6 by the gadget, which is calculated by $s = (-d) \bmod 6$. In the case of the U-turn gadget $s = (-77) \bmod 6 = 1$.

Table 7.1   Net delay and phase shift of the six turn gadgets [33].

Gadget	Net delay d	Phase shift s
60°-right-turn	5	+1
120°-right-turn by 2 blocks	304	+2
120°-right-turn by 3 blocks	220	+2
120°-right-turn by 5 blocks	178	+2
Backward-turn	73	+5
U-turn	77	+1

Combining the six turn gadgets, we can not only change the moving direction of a glider-6 freely, but also change the phase of it. For example, placing the backward turn gadget on the bidirectional signal path in Fig. 7.7 appropriately, we have another 60°-right-turn gadget, which performs two 120°-right-turns and one backward turn. Therefore, the phase shift value in this case is $(2 + 2 + 5) \bmod 6 = 3$, which differs from the phase shift value 1 of the 60°-right-turn gadget in Fig. 7.5. In such a way, the phase of a glider is adjusted.

Figure 7.10 shows operations of shifting a fin by colliding a glider-6. Thus, a fin can be used as a *position marker* that works as a kind of memory. A fin is pushed by colliding a glider-6 as in Fig. 7.10(a). In this case, the position is shifted by $(3, -1)$ (see the coordinates of a TPCA space given in Fig. 2.15). A fin is pulled by colliding a glider-6 as in Fig. 7.10(b), where the position is shifted by $(-3, -1)$. It is called a type-1 pulling operation. In each of these two cases, the glider-6 moves backward after the collision. There is another type of pulling as in Fig. 7.10(c). In this case, the fin is shifted by $(-6, 0)$. It is called a type-2 pulling operation. After shifting the fin, the glider-6 continues to move to the same direction. The three mechanisms in Fig. 7.10 are called *shifting gadgets*.

In each of these shifting operations, the phase of a fin does not change, *i.e.*, the patterns at $t \equiv 0 \pmod 6$ are always the same. This property makes it easy to use a fin as a position marker. On the other hand, the phase of a glider-6 changes after a shifting process. It can be adjusted by using turn gadgets properly (see Table 7.1).

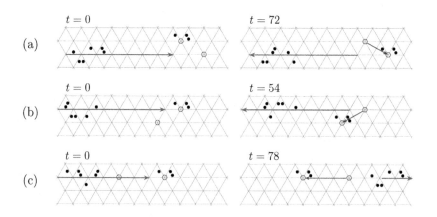

Fig. 7.10   Shifting a fin by colliding a glider-6 in ETPCA-0347. (a) Pushing [66], (b) type-1 pulling [66], and (c) type-2 pulling.

7.1.3 *Making RLEMs 4-31 and 2-17 in ETPCA-0347*

Here, we construct patterns that simulate RLEMs in ETPCA-0347. In Chap. 6 composing methods of an RE in reversible ESPCAs are shown. On the other hand, in ETPCA-0347, we first make RLEM 4-31 (Fig. 7.11). The reason is that it is easier to construct it than an RE, since there are only two cases of state-input pairs that cause state-changes. They are the pairs $(0, d)$ and $(1, b)$ (see Fig. 7.11). Due to this property, we can compose RLEM 4-31 using one position marker. Note that, in ESPCAs 01c5ef and 01caef, two and four position markers are required to make an RE (Secs. 6.1 and 6.2), since there are four cases of state changes. As we have seen in Sec. 4.3.5.2, RTMs can be constructed using RLEM 4-31 easily.

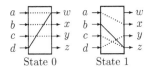

Fig. 7.11   RLEM 4-31.

Figure 7.12 is a pattern that simulates RLEM 4-31 in ETPCA-0347. Small circles around the middle of the pattern show possible positions of a fin, a position marker. If it is at the lower (upper, respectively) position, it is regarded to be the state 0 (state 1).

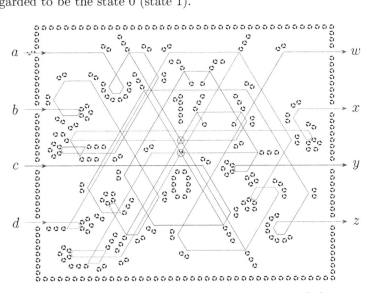

Fig. 7.12   RLEM 4-31 implemented in ETPCA-0347 [66].

In Fig. 7.12, all the trajectories of signals for the eight cases of state-input pairs are written, and thus they are complicated. Since there are many access paths to the position marker to know and/or change the state, it is hard to understand them. Below we consider three cases, and explain how the pattern works in each case. Other cases are similar.

We assume that a glider-6 given at an input port is of phase 0 when $t \equiv 0 \pmod 6$. Then, at the time the glider-6 reaches an output port, it is adjusted so that it is of phase 0 when $t \equiv 0 \pmod 6$.

Figure 7.13 shows the case $(0, a)$. The glider-6 from the input port a first goes to the point P, and then Q. At Q, the glider-6 moves straight ahead without interacting with the position marker, since the state of the RLEM is 0. Finally, it goes out from the port w. By above, $(0, a) \mapsto (0, w)$ is performed.

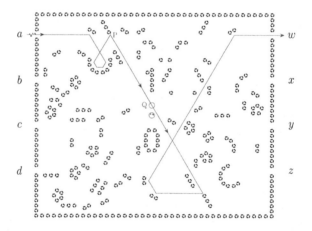

Fig. 7.13   RLEM 4-31 in ETPCA-0347. The trajectory of the glider shows the case where the state-input pair is $(0, a)$.

Figure 7.14 is the case $(1, a)$. The glider-6 first accesses the position marker at Q moving from P. At Q, the trajectory branches from that of the case $(0, a)$. Here, by the type-1 pulling operation of Fig. 7.10(b), the marker moves to the position indicated by a very small circle which is to the left of Q, and then the glider-6 goes backward. Next, the glider-6 accesses the marker from R. It makes the marker go back to the position Q by the pushing operation of Fig. 7.10(a). By this, the original state 1 of the RLEM 4-31 is restored. Finally, the glider-6 goes out from the port x. By above, $(1, a) \mapsto (1, x)$ is performed.

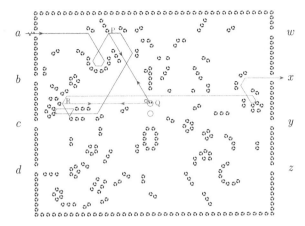

Fig. 7.14   RLEM 4-31 in ETPCA-0347. The trajectory shows the case of $(1, a)$.

Figure 7.15 is the case $(0, d)$. The glider-6 accesses the position marker at S moving from T. By the type-1 pulling operation of Fig. 7.10(b), the marker moves to the position of a very small circle between Q and S. Then, the glider-6 goes to U, and pushes the marker by the operation of Fig. 7.10(a). By this, the marker moves to the position Q. Note that, in this case, both pushing and type-1 pulling operations are used so that the marker positions for 0 and 1 keep enough space. Finally, the glider-6 goes out from the port w. The trajectories of the cases $(0, a)$ and $(0, d)$ are reversibly merged at the point Q, and they lead to the port w. Hence, we have $(0, d) \mapsto (1, w)$.

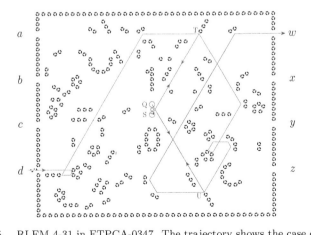

Fig. 7.15   RLEM 4-31 in ETPCA-0347. The trajectory shows the case of $(0, d)$.

Next, we compose RLEM 2-17 (Fig. 7.16) in ETPCA-0347. Though there are four cases of state-input pairs that cause state changes, the total number of cases is also four. Because of this property, RLEM 2-17 can be implemented in ETPCA-0347 by only one position marker as shown in Fig. 7.17. It is known that RLEM 2-17 is intrinsically universal (Theorem 4.11). Therefore, any RTM is composed only of it. However, resulting circuits are more complex than to use RLEM 4-31.

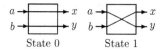

Fig. 7.16   RLEM 2-17.

In Fig. 7.17, two small circles indicate possible positions of a position marker (*i.e.*, a fin). If it is at the lower (upper, respectively) position, we regard it is in the state 0 (state 1).

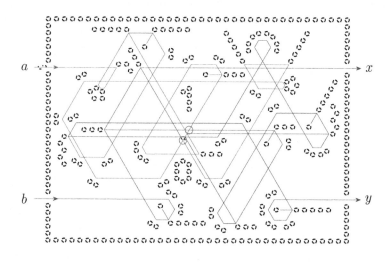

Fig. 7.17   RLEM 2-17 implemented in ETPCA-0347.

Figure 7.18 shows the case where the state-input pair is $(0, a)$. The signal (*i.e.*, a glider) first goes to the point P, and then to Q. It moves straight ahead at Q. Thus, the signal knows that the state is 0, and goes from Q to S. By the type-2 pulling operation shown in Fig. 7.10(c), the position marker is shifted from S to Q. Finally, the signal goes out from the port x. By above, $(0, a) \mapsto (1, x)$ is performed.

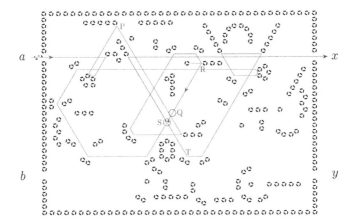

Fig. 7.18   RLEM 2-17 in ETPCA-0347. The trajectory shows the case of $(0, a)$.

Next, consider the case where the state-input pair is $(1, b)$. Figure 7.18 shows its simulating process. The signal first accesses the position marker by moving from U to Q. By this, the marker is shifted to the position indicated by a very small circle using the type-1 pulling operation. Next, moving from V to S, the signal accesses the position marker. By this, the marker is pushed to the point S. Note that the trajectories of the cases $(0, a)$ and $(1, b)$ are reversibly merged at the point S, and the both trajectories lead to the port x. Thus, $(1, b) \mapsto (0, x)$ is performed. Remaining cases $(0, b)$ and $(1, a)$ are performed in a similar manner.

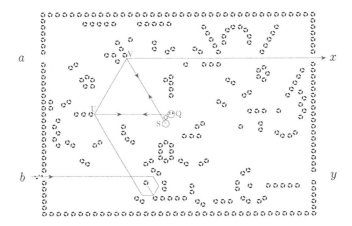

Fig. 7.19   RLEM 2-17 in ETPCA-0347. The trajectory shows the case of $(1, b)$.

7.1.4 *Making RTMs by RLEMs in ETPCA-0347*

A construction method of RTMs using RLEM 4-31 is given in Sec. 4.3.5.2. Replacing each occurrence of RLEM 4-31 in Fig. 4.34 by the pattern in Fig. 7.12, and connecting the patterns appropriately, we have a configuration that simulates RTM T_{parity} in ETPCA-0347. Figure 7.20 shows the configuration simulated on Golly. Any RTM can be constructed by this method. Therefore, ETPCA-0347 is Turing universal.

Fig. 7.20   RTM T_{parity} in ETPCA-0347 made of RLEM 4-31 [21]. It simulates the circuit of Fig. 4.34. Each small rectangle is the pattern of RLEM 4-31 given in Fig. 7.12.

Construction of RTMs in ETPCA-0347 using RLEM 2-17 is much more complex than the case of using RLEM 4-31. Figure 7.21 gives a pattern of a tape cell module (see Sec. 4.3.5) implemented in ETPCA-0347 using RLEM 2-17. It is based on the idea of M. Cook and E. Palmiere (Theorem 4.11). A state module is also implemented by a similar technique. Therefore, any RTM is constructed in ETPCA-0347 systematically.

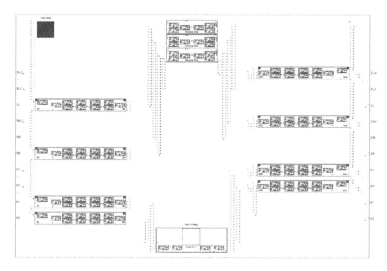

Fig. 7.21   Tape cell module realized in ETPCA-0347 composed of RLEM 2-17 [21]. Each small rectangle is the pattern of RLEM 2-17 given in Fig. 7.17.

7.1.5 *Making a glider gun and an absorber*

We compose a glider gun and a glider absorber in ETPCA-0347. Though they are not related to the universality of ETPCA-0347, they are constructed by utilizing the useful phenomena shown in Sec. 7.1.2, and the processes of changing the number of glider-6's given below. Therefore, we describe them here.

Colliding two glider-6's as in Fig. 7.22, we have three gliders. To the contrary, colliding three glider-6's as in Fig. 7.23, we have two. Thus, we can increase and decrease the number of gliders by these processes. Note that the two processes are T-symmetric each other as discussed in Fig. 3.29 of Sec. 3.3.3. In fact, the configuration at $t = 0$ of Fig. 7.23 is obtained by applying the involution $H^{\mathrm{rev}} \circ H^{\mathrm{refl}}$ to the configuration at $t = 30$ of Fig. 7.22.

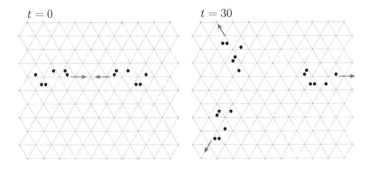

Fig. 7.22　Generating three glider-6's from two in ETPCA-0347 [33].

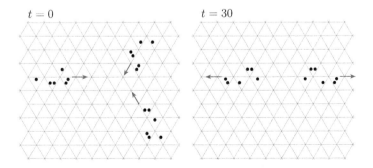

Fig. 7.23　Generating two glider-6's from three in ETPCA-0347 [33].

By the process of Fig. 7.22, we can design a *glider gun* as shown in the left part of Fig. 7.24. There, one of the three gliders generated by a collision of two gliders is emitted to the north-west direction. The remaining two gliders circulate in the gun with the period of 834, and are used to generate the next three gliders. Many blocks are used to form 60° and 120°-right-turns, a backward-turn, and a U-turn to adjust the period and the phase of gliders. By this, a glider stream is obtained from the output port at the north-west position with the period of 834 steps.

Different from the glider guns given in Sec. 5.5.2, this type of glider gun has a somewhat "programmable" nature. The gun in Fig. 5.80 has a fixed emission period that cannot be altered. On the other hand, by changing the circulation period of two gliders in the gun of Fig. 7.24, the emission period can be varied.

Symmetrically, by the process of Fig. 7.23, a *glider absorber* is obtained as shown in the right part of Fig. 7.24. There, an incoming glider from the north-east direction is reversibly erased by the two gliders, which circulate in the absorber with the period 834. Therefore, an input glider of a right phase must also be given every 834 steps. Otherwise, the glider absorber will be broken.

Note that the glider gun and the absorber themselves are T-symmetric each other. Namely, the glider absorber is obtained by applying the involution $H^{\mathrm{rev}} \circ H^{\mathrm{refl}}$ to the glider gun. This type of gun and absorber in ETPCA-0347 was first proposed in [33], and revised in [64].

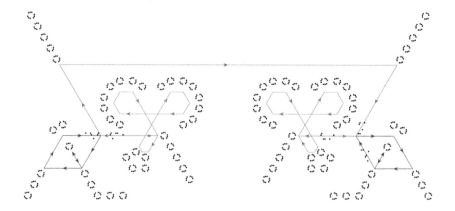

Fig. 7.24   Glider gun (left) and absorber (right) [64]. Gliders are generated by the gun every 834 steps, which form a glider stream. In this figure, the glider stream is supplied to the absorber, and gliders are reversibly erased by it.

7.2 ETPCA-034z, a partial ETPCA

A *partial ETPCA* is a variant of an ETPCA whose local function is partially defined. ETPCA-034z is an example of a partial ETPCA. The set of its local transition rules is shown in Fig. 7.25, where the fourth local transition rule is missing. Its local function f_{034z} is thus

$$f_{034z}(0,0,0) = (0,0,0), \quad f_{034z}(0,1,0) = (0,1,1), \quad f_{034z}(1,0,1) = (1,0,0),$$

but $f_{034z}(1,1,1)$ is undefined. Note that the undefined local transition rule is represented by the variable z in its identification number.

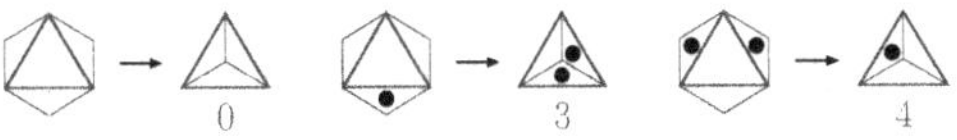

Fig. 7.25　Three local transition rules of ETPCA-034z.

The global function F of a partial ETPCA is similarly defined as in Definition 2.21 based on its partial local function f. Different point is that, for a configuration α, if there exists $\mathbf{x} \in \mathbb{Z}^2$ such that $F(\alpha)(\mathbf{x})$ is undefined, then we say $F(\alpha)$ is undefined.

Consider a partial ETPCA having a partial global function F. Its configuration α is called *valid*, if $F^t(\alpha)$ is defined for all $t > 0$. Otherwise, α is called an *invalid configuration*.

For example, look at the configuration of ETPCA-0347 at $t = 30$ in Fig. 7.9. It is invalid in ETPCA-034z. This is because the 13-th cell in the second row has the state 1's in all of the three neighboring parts, but $f_{034z}(1,1,1)$ is undefined. Hence, the configuration at $t = 0$ is also invalid. Invalid configurations cannot be used in the partial ETPCA.

On the other hand, we can verify that configurations shown in Figs. 7.2–7.4 are all valid in ETPCA-034z. In fact, these configurations do not use the fourth local transition rule of ETPCA-0347 when they evolve.

7.2.1 *Useful patterns and phenomena in ETPCA-034z*

As it is seen above, a glider (Fig. 7.2), a fin (Fig. 7.3) and a block (Fig. 7.4) are valid configurations in ETPCA-034z, and thus usable as a basic patterns. However, most phenomena shown in Sec. 7.1.2 are not usable in ETPCA-034z, since the fourth local transition rule of ETPCA-0347 is used in their evolution processes (we checked them using the CA simulator Golly). Only the configurations contained in the 60°-right-turn process

(Fig. 7.5) are valid ones. Therefore, in the following, we use the 60°-right-turn gadget to control the moving direction of a glider-6.

The three shifting gadgets for a position marker shown in Fig. 7.10 are also not usable in ETPCA-034z. Thus, we have to use another type of shifting gadget shown in Fig. 7.26. It performs a type-3 pulling operation. By this, the position of a fin is shifted by $(-3, 1)$, and the glider-6 makes a 60°-left-turn. This evolution process contains only valid configurations of ETPCA-034z (it is checked by using the CA simulator Golly). Similar to the three cases shown in Fig. 7.10, the phase of a fin does not change (see $t = 126$ in Fig. 7.26). On the other hand, the phase of a glider-6 shifts by 3 (see $t = 123$ in Fig. 7.26)). Thus, we have to develop a technique to adjust its phase. It is explained in the next section.

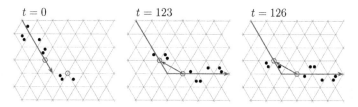

Fig. 7.26   Type-3 pulling operation of a fin by colliding a glider-6 in ETPCA-034z.

In ETPCA-034z, we use three basic patterns, which are a *glider-6*, a *fin* and a *block*, and two gadgets, which are the *60°-right-turn gadget* (Fig. 7.5) and the type-3 *shifting gadget* for a fin (Fig. 7.26).

7.2.2   *Making RLEM 2-2 in ETPCA-034z*

Here, we design a pattern that simulates RLEM 2-2 (Fig. 7.27) in ETPCA-034z. Though RLEM 2-2 is not intrinsically universal (Theorem 4.10), M. Cook and E. Palmiere showed that it is Turing universal (Theorem 4.12), *i.e.*, any RTM can be composed only of it (its construction is quite complex). Therefore, ETPCA-034z is also Turing universal, if RLEM 2-2 is realized in it. Note that, so far, it is not known whether a 2-state RLEM other than RLEM 2-2 can be simulated in ETPCA-034z.

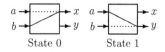

Fig. 7.27   RLEM 2-2.

Since we can use only one kind of a signal turn operation, a difficulty arises in adjusting the phase of a glider-6. In particular, there is a limitation when we shift a position marker successively. Figure 7.28 shows that after a north-west (NW) shifting of a position marker, a north (N) shifting is possible. We can see that, in a similar manner, a south shifting (S) is also possible after an NW shifting. However, after NW, neither north-west (NW), north-east (NE), south-west (SW), nor south-east (SE) shifting is possible, because the phase of a glider-6 is not adjustable by the 60°-right-turn operation. Likewise, after an N or S shifting, only NE or SW shifting is possible. Thus, a legal shifting sequence must satisfy the following.

$$\text{NW or SE} \rightarrow \text{N or S} \rightarrow \text{NE or SW} \rightarrow \text{NW or SE}$$

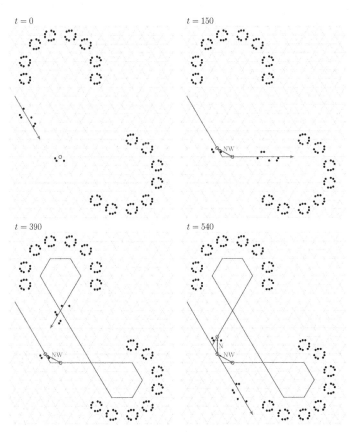

Fig. 7.28   Two successive shifts of a position marker in ETPCA-034z. After a north-west (NW) shifting, a north (N) shifting is possible.

To keep a state of RLEM 2-2, we use a marker "A" shown in Fig. 7.29. There are ten possible positions of the marker A. They are A_0, ..., A_9. Among them A_0 and A_5 correspond to the states 0 and 1 of RLEM 2-2, respectively. The other positions are transient ones. Giving a glider-6 of a right phase from the port G_i as a signal, a position marker at A_i is shifted to the point A_{i+1}, where $i \in \{0, \ldots, 9\}$ and $+$ is the mod 10 addition. Finally, the signal goes out from G'_i. Changing the state from 0 to 1 is done by giving signals from G_0, G_1, G_2, G_3 and G_4 in this order (note that how to give such a sequence is explained below).

Testing the state of the RLEM 2-2 is performed by a signal from G_0. If it is in the state 0, then the signal goes out from G'_0, and the position marker is shifted to A_1. If it is in the state 1, then the signal moves straight ahead, and goes out from G''_0. By this, the signal path branches. Testing is also performed from the input G_5 in a similar way.

Merging of two signal paths is also possible. A signal from G_4 that shifts a marker from A_4 to A_5 will go out from the port G'_4, changing the state to 1. On the other hand, if a signal is given to G_{M_0} in the state 0, it also goes out from G'_4. This is a reversible merge that can be used for implementing RLEM 2-2. Similarly, the paths from G_9 and G_{M_1} are reversibly merged.

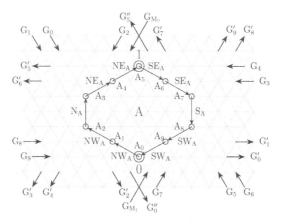

Fig. 7.29   Possible positions of a marker A for memorizing a state of RLEM 2-2.

We explain how the state transition of RLEM 2-2 is performed. To change the state from 0 to 1 in Fig. 7.29, a position marker A should be shifted to the directions NW_A, NW_A, N_A, NE_A, NE_A, in this order. However, there are illegal successive shifts, *e.g.*, NW_A, NW_A. To avoid such an illegal pair, we add two *ancillary position markers* B and C shown in Fig. 7.30. If the position marker B (C, respectively) is at B_0 (C_0), we regard the state is 0. If the marker is at B_5 (C_5), the state is 1.

The following shift sequence changes the state from 0 to 1.

$$NW_A, N_B, SW_C, NW_A, N_B, SW_C, NW_C, N_A, NE_B,$$
$$SE_B, N_C, NE_A, SE_B, N_C, NE_A$$

We can see that it is a legal shifting sequence. Furthermore, if we take a subsequence related to the position marker A (or B, C, respectively), then the subsequence makes the marker go from A_0 (B_0, C_0) to A_5 (B_5, C_5).

By reversing the direction of each shift in the above sequence, we have a shift sequence that changes the state from 1 to 0.

$$SE_A, S_B, NE_C, SE_A, S_B, NE_C, SE_C, S_A, SW_B,$$
$$NW_B, S_C, SW_A, NW_B, S_C, SW_A$$

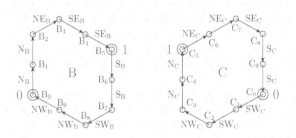

Fig. 7.30   Ancillary position markers B and C for adjusting the phase of a signal.

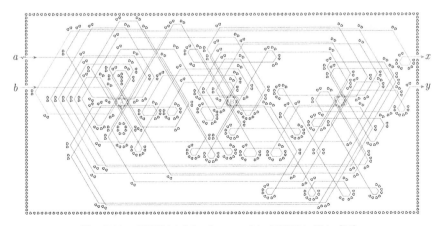

Fig. 7.31   RLEM 2-2 implemented in ETPCA-034z [21].

Figure 7.31 shows a pattern that simulates RLRM 2-2 in ETPCA-034z. Below, the two cases, where the state-input pairs are $(0, a)$ and $(0, b)$, are explained. Other cases are similar to these two.

Figure 7.32 is the case $(0, a)$. The input signal first goes to G_5 in Fig. 7.29, and tests if the state is 0 or 1. Since it is 0, the signal then goes to G_{M_0} to merge with another path that leads to the same output port. It finally comes out from the port x without changing the state.

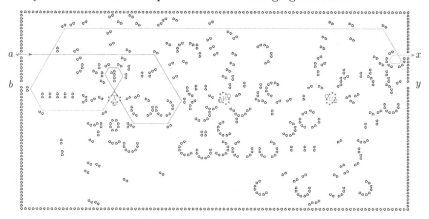

Fig. 7.32   RLEM 2-2 in ETPCA-034z. The trajectory shows the case of $(0, a)$.

Figure 7.33 is the case $(0, b)$. The input signal first goes to G_0, and tests if the state is 0 or 1. Since it is 0, the signal executes the shift sequence that changes the state from 0 to 1. After that, it goes to G_{M_0} to merge with the path of the case $(0, a)$, and comes out from x.

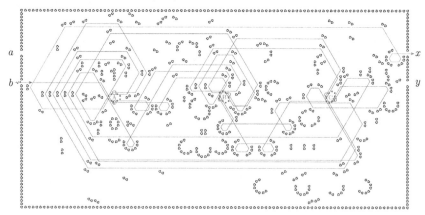

Fig. 7.33   RLEM 2-2 in ETPCA-034z. The trajectory shows the case of $(0, b)$.

7.3 ETPCA-0157

Figure 7.34 gives the local function of ETPCA-0157. It is reversible and conservative. In [30], it was shown that a Fredkin gate, a universal reversible logic gate, is implemented in it. Here, we make RLEM 4-31 in it.

As it was seen in Sec. 5.3.2.1, various kinds of space-moving patterns exist in ETPCA-0157. However, those patterns so far found have very long periods. Hence, it is difficult to control their moving directions and phases. Instead, we use here a single particle that can move along a "wall," a kind of transmission wire, as an information carrier. Thus the basic method of processing a signal is different from the case of ETPCA-0347.

Fig. 7.34   Local function of ETPCA-0157.

7.3.1 *Useful patterns and phenomena in ETPCA-0157*

We use two kinds of basic patterns. The first one is a *rotor*, consisting of one particle shown in Fig. 7.35. If it exists solely, it rotates with the period of 6. As we shall see below, it can move along a wire. Hence, it is used as a *signal*. As explained later, an input/output signals and a state signals in ETPCA-0157 consist of five consecutive signals that move along a wire.

Fig. 7.35   Rotor, a periodic pattern, in ETPCA-0157. It is used as a signal.

The second is a *block*, a stable pattern composed of 12 particles shown in Fig. 7.36. It is used to make signal transmission wires.

Fig. 7.36   Block, a stable pattern in ETPCA-0157.

There are several kinds of useful phenomena for processing signals. Figure 7.37 shows that a signal can move along a sequence of blocks or compound blocks. Thus, they are used as *transmission wires*. Figure 7.37(a) is a type-1 horizontal transmission wire. There, it takes 12 steps for a signal to move to the next block. Figure 7.37(b) is a type-2 transmission wire, where a signal moves to the next block in 4 steps.

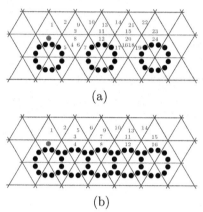

(a)

(b)

Fig. 7.37   (a) Type-1, and (b) type-2 horizontal transmission wire for a right-moving signal. The numbers in the figure show the positions of the signal at time t, where $1 \leq t \leq 24$ in (a), and $1 \leq t \leq 16$ in (b). A leftward signal can move beneath the wires.

Figure 7.38 shows vertical transmission wires. In the type-1 transmission wire (Fig. 7.38(a)), it takes 12 steps for a signal to move to the next block, while in the type-2 wire (Fig. 7.38(b)), it takes 8 steps to move to the next block. A horizontal wire and a vertical wire can be connected easily. Furthermore, a wire can be bent relatively freely.

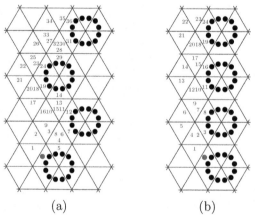

(a)   (b)

Fig. 7.38   (a) Type-1 vertical transmission wire, and (b) type-2 wire for an upward-moving signal. A downward signal can travel along the right-side of the wires.

Phenomena of interacting two signals caused by the third local transition rule (Fig. 7.34) can be used for crossing signals in the 2-dimensional cellular space. If a space-moving pattern (glider) is used as a signal as in ETPCA-0347, then no signal crossing module is required. However, in ETPCA-0157, since a signal moves along a wire, such a crossing device is necessary.

Fig. 7.39 Signal-crossing module for a rightward-moving signal and an upward-moving signal in ETPCA-0157. This figure shows how an upward signal given at the bottom block moves. It reaches the top block at $t = 144$ as if it goes along a type-1 vertical wire.

Figure 7.39 is a *signal-crossing module* for rightward and upward-moving signals. Note that crossing modules for signals moving to other directions, *e.g.*, rightward and downward, are similarly designed. In this pattern, there are two signals that are rotating with a period of 6 near the crossing point of input signals. An input signal from the west or the south will interact with these two signals, and its moving direction is controlled.

Figure 7.39 shows the process where a signal goes from the south to the north, while Fig. 7.40 shows the one where a signal goes from the west to the east. Figure 7.41 gives the details of the interaction of signals in Fig. 7.40. At $t = 22$ the input signal meets one of the control signals, and its trajectory is changed at $t = 23$. Then, it again interacts with the second control signal at $t = 23$. By this, the input signal is directed to the upper-right block as seen in $t = 24, \ldots, 27$. After that, the signal travels rightward along a type-1 horizontal wire.

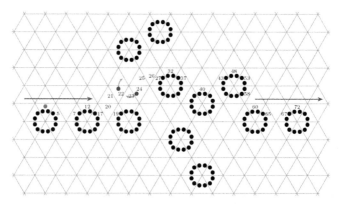

Fig. 7.40 This figure is a middle part of the signal-crossing module given in Fig. 7.39. It shows how a rightward-moving signal travels. It reaches the rightmost block at $t = 72$ as if it moves along a type-1 horizontal wire.

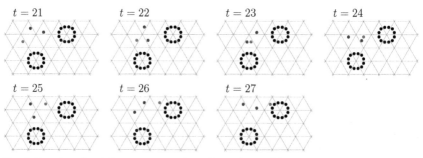

Fig. 7.41 Details of the movement of a rightward signal from $t = 21$ to 27 in Fig. 7.40.

As seen above, the period of a crossing module is 6, since it contains two rotating signals. Therefore, when connecting such modules with wires, it is convenient to use only type-1 wires (Figs. 7.37(a) and 7.38(a)). This is because the total delay of a signal by such a wire is a multiple of 12. Hence, there is no need to adjust the phase of a signal. Thus, we use the type-1 wires as the standard ones, though the signal speeds of type-2 wires are faster than those of type-1 wires. Note that, however, inside the crossing module, several segments of type-2 wires and their rotated ones are used to adjust the timing. It is also the case in the modules given below.

7.3.2 *Making RLEM 4-31 in ETPCA-0157*

Here, we design RLEM 4-31 in ETPCA-0157. To do so, we first give a Y-module and a bridge. Then, a flip-flop module is composed out of them. Finally, RLEM 4-31 is constructed using a flip-flop.

A *Y-module* is a pattern composed of six blocks shown in Fig. 7.42. This module is for keeping the state of a flip-flop memory. A signal can rotate around it with a period of 150 steps. We can put several evenly spaced signals to reduce the period of the module. When designing a module, its period p should be carefully chosen. Let $\mathrm{lcm}(x, y)$ denotes the least common multiple of integers x and y. Since we use type-1 wires as the standard ones, and their delay is a multiple of 12, the period p should be chosen so that $\mathrm{lcm}(p, 12)$ becomes as small as possible. If we put 25 signals on the Y-module, then $p = 6$ and thus $\mathrm{lcm}(p, 12) = 12$. However, it does not work, since adjacent two signals interfere on a wire. Here, we put five signals on the Y-module. These five signals on the Y-module is called a *state signals* In this case the Y-module has the period $p = 30$, and thus $\mathrm{lcm}(p, 12) = 60$. As we shall see below, the periods of a flip-flop and RLEM 4-31 are also 30. Therefore, connecting these modules by type-1 wires whose lengths are of multiples of 5 blocks, each module receives a signal at a right timing.

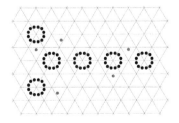

Fig. 7.42   Y-module in ETPCA-0157. Five state signals are moving around it.

A flip flop has two Y-modules as in Fig. 7.46. If five state signals are rotating on the left (right, respectively) Y-module, then we regard it is in the state 0 (state 1). To move the state signals from the left to the right, or its inverse, we use a *bridge* as an intermediate module to adjust the phase of the state signals. A transfer process from a Y-module to a bridge is shown in Fig. 7.43. It is done by giving five control signals from the north. How a state signal is shifted to the bridge by a control signal is shown in Fig. 7.44.

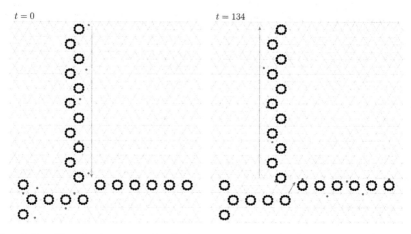

Fig. 7.43　Transferring state signals from a Y-module (left) to a bridge (right) in ETPCA-0157 by giving control signals from the north.

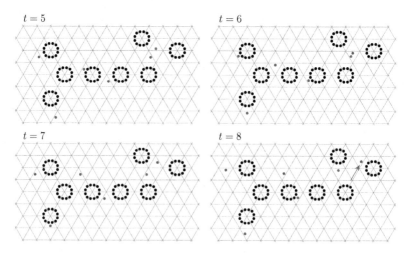

Fig. 7.44　Details of the transfer process of Fig. 7.43.

Two bridge modules are placed between the two Y-modules as in Fig. 7.45. We explain below why two bridges are necessary. Consider the following process. At first, state signals are on the left Y-module. Giving control signals by the method of Fig. 7.43, the state signals are transferred to the left bridge. After rotating around the left bridge a few times, the state signal are shifted to the right bridge by giving control signals from the north. Likewise, after rotating around the right bridge a few times, the state signal are shifted to the right Y-module. The state signals rotate around the right Y-module many times. Then, by giving control signals from the south, the state signals move to the right bridge, then to the left bridge, and finally reach the left Y-module.

Let delay(P, Q) denote the minimum time for a signal to reach the point Q from P (note that control signals like the ones in Fig. 7.43 are supplied if needed). Then, we can calculate the following easily.

delay(A, B) = delay(C, D) = delay(E, F) = 10
delay(G, H) = delay(I, J) = delay(K, L) = 10
delay(B, C) = delay(D, E) = delay(H, I) = delay(J, K) = 60
delay(F, G) = delay(L, A) = 135

Assume a state signal starts the point A at $t = 0$, and returns to A at $t = t_1$ in the above process. Then, t_1 is as follows, where k is a nonnegative integer. Note that the time to go around a bridge just once is 150 steps.

$$t_1 = 10 \times 6 + 60 \times 4 + 135 \times 2 + 150k$$

We can see that t_1 is divisible by 30, and thus the state signals can be moved to the right again by the control signals of the same phase as those of $t = 0$. It makes it easy to use the module.

If there is no bridge, the returning time t_2 of the state signal is as follows.

$$t_2 = 10 \times 2 + 135 \times 2 + 150k$$

Since t_2 is indivisible by 30, to move the state signals again, the phase of the control signals must be shifted. The case of one bridge is also similar.

Fig. 7.45   Two bridge modules between the Y-modules in ETPCA-0157.

Figure 7.46 is a pattern of a *flip-flop* module. Assume it is in the state 0, *i.e.*, state signals go around the left Y-module. If five signals are given to the S input port with a right timing, then the state signals are first moved to the left bridge, then to the right bridge, and finally to the right Y-module. Namely, the state is set to 1. The input signals goes out from the output port S'. Likewise, if five signals are given to R when the state is 1, then it is reset to 0, and the signals go out from the port R'.

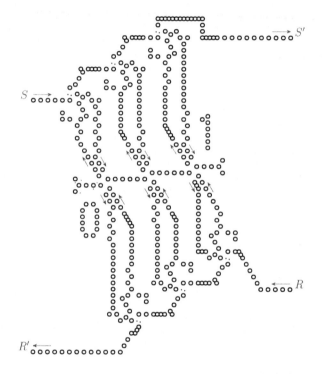

Fig. 7.46   Flip-flop module in ETPCA-0157.

To construct RLEM 4-31 by extending a flip-flop module, we need a mechanism that performs branching and merging of signal paths by testing if state signals exist or not on a Y-module. An implementation method of such a mechanism is shown in Fig. 7.47.

Giving test signals from the path A, branching is performed. If state signals exist on the Y-module, then the test signals goes to the path B (Fig. 7.47(a)). On the other hand, if no state signal exists, then the test signals goes to the path C (Fig. 7.47(b)).

The process of Fig. 7.47(a) is also used for merging two signal paths. If signals are given from the path D in Fig. 7.47(c) under the assumption that no state signal exists on the Y-module, then they goes to the path B. By this, a reversible merge of the paths A and D into B is realized.

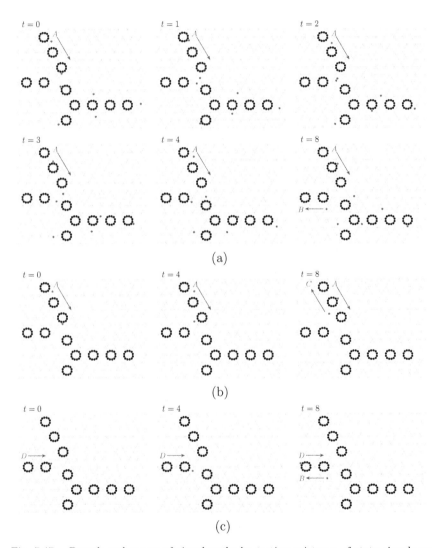

Fig. 7.47   Branch and merge of signal paths by testing existence of state signals on a Y-module in ETPCA-0157. (a) Sensing that state signals exist. It is used for both branching and merging. (b) Sensing that no state signal exists. It is for branching. (c) Sensing that no state signal exists. It is for merging.

Figure 7.48 is a pattern that simulates RLEM 4-31 in ETPCA-0157. It has a flip-flop module in its center part. Around it, four sets of branch-and-merge mechanisms are attached.

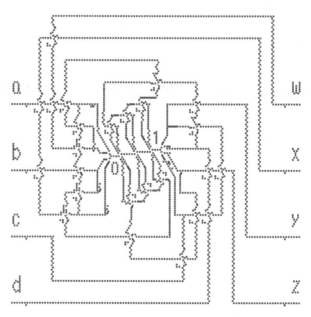

Fig. 7.48   RLEM 4-31 implemented in ETPCA-0157.

7.3.3   *Making RTMs by RLEM 4-31 in ETPCA-0157*

Figure 7.49 is a configuration of RTM T_{parity} implemented in ETPCA-0157. It is obtained by replacing each occurrence of RLEM 4-31 in Fig. 4.34 by the pattern of Fig. 7.48 and connecting the patterns appropriately. Since any RTM can be constructed in this way, ETPCA-0157 is Turing universal.

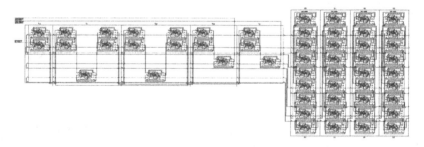

Fig. 7.49   RTM T_{parity} in ETPCA-0157 simulated on Golly [21].

7.4 ETPCA-013z, a partial ETPCA

Figure 7.50 shows the partial local function of ETPCA-013z. Similar to ETPCA-034z (Sec. 7.2), it has only three local transition rules, and thus it is extremely simple. In ETPCA-013z, however, RLEM 4-31, an intrinsically universal RLEM, can be implemented using only the three local transition rules.

Fig. 7.50 Local transition rules that define the partial local function of ETPCA-013z.

7.4.1 Useful patterns and phenomena in ETPCA-013z

Like the case of ETPCA-0157, there are various kinds of space-moving patterns in ETPCA-0137 (see Sec. 5.3.2.1). We can see that the patterns shown in Fig. 5.58 also work as space-moving patterns in ETPCA-013z, *i.e.*, they use only the three local transition rules (it has been verified by a computer simulator). However, since their periods are very long, it is difficult to use them as signals. Thus, we take an approach similar to ETPCA-0157 for constructing RLEM 4-31.

We use a single particle as a *signal* (Fig. 7.51) in ETPCA-013z. It is exactly the same as in ETPCA-0157.

Fig. 7.51 Signal in ETPCA-013z. If a signal exists solely, it rotates with a period 6.

We also use a *block* shown in Fig. 7.52. It is for composing signal transmission wires. The main reason that RLEM 4-31 is realizable in ETPCA-013z is that the block has six particles rather than 12. By this, signal transmission processes are performed using only three local transition rules. Note that the 12-particle block in ETPCA-0157 (Fig. 7.36) becomes a periodic pattern of period 7 in ETPCA-013z, and is not suited for a wire.

$t = 0$

Fig. 7.52 Block, a stable pattern in ETPCA-013z.

A signal can move along a sequence of blocks. Figure 7.53 is a horizontal *transmission wire* in ETPCA-013z. It takes 12 steps for a signal to move by one block.

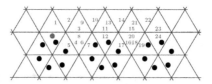

Fig. 7.53   Horizontal transmission wire for a right-moving signal in ETPCA-013z [63]. The numbers in the figure show the positions of the signal at time t, where $1 \leq t \leq 24$. A leftward signal can move beneath the wires.

Figure 7.54 shows vertical transmission wires. Like the case of ETPCA-0157 (Fig. 7.38), in the type-1 transmission wire (Fig. 7.54(a)), it takes 12 steps for a signal to reach the next block, while in the type-2 wire (Fig. 7.54(b)), it takes 8 steps to reach the next block.

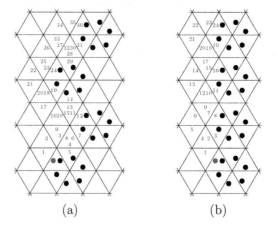

(a)   (b)

Fig. 7.54   (a) Type-1, and (b) type-2 vertical transmission wires for an upward-moving signal in ETPCA-013z [63]. A downward signal can move along the right-side of the wires.

A *signal-crossing module* is realized by utilizing phenomena of interacting two signals. Figure 7.55 is a crossing module for right-moving and upward-moving signals. Modules for other signal directions are similarly composed. Besides many blocks, two control signals, which rotate with a period of 6, are contained near the crossing point of the module. The control signal at the north-east (south-west, respectively) position is for interacting with an upward-moving (rightward-moving) signal. In Fig. 7.55, the upward signal interacts with the north-east control signal at $t = 67$, and continues to move upward.

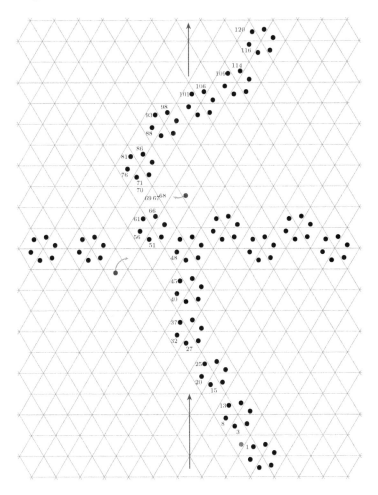

Fig. 7.55   Signal-crossing module for a rightward signal and an upward signal in ETPCA-013z [63]. This figure shows how an upward signal given at the bottom block moves. It reaches the top block at $t = 120$ as if it goes along a type-1 vertical wire.

Figure 7.56 shows the trajectory of a right-moving signal. An input signal interacts with the south-west control signal at $t = 19$, and then continues to move rightward. Details of the interaction process is shown in Fig. 7.57.

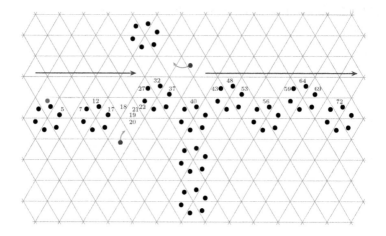

Fig. 7.56   This figure is a middle part of the signal-crossing module given in Fig. 7.55. It shows how a rightward-moving signal travels. It reaches the rightmost block at $t = 72$ as if it moves along a straight horizontal wire of Fig. 7.53.

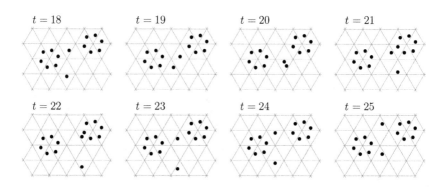

Fig. 7.57   Details of the movement of a rightward signal from $t = 18$ to 25 in Fig. 7.56.

Note that, since the period of a signal-crossing module is 6, it is convenient to use a horizontal wire (Fig. 7.53) and a type-1 vertical wire (Fig. 7.53(a)) as standard wires, since a signal moves to the next block in 12 steps on these wires. However, in a crossing module itself and in modules composed below, segments of a type-2 vertical wire and rotated wires are also used for adjusting signal timings.

7.4.2 *Making RLEM 4-31 in ETPCA-013z*

We employ a method similar to the case of ETPCA-0157 for composing RLEM 4-31. Namely, we use a Y-module to keep the state of RLEM 4-31. Figure 7.58 is the *Y-module* in ETPCA-013z. A signal can move around it with a period of 150. As in the case of ETPCA-0157, we put five equally spaced signals as *state signals*, to reduce its period to 30.

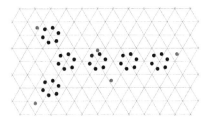

Fig. 7.58   Y-module in ETPCA-013z [63]. Five state signals are moving around it.

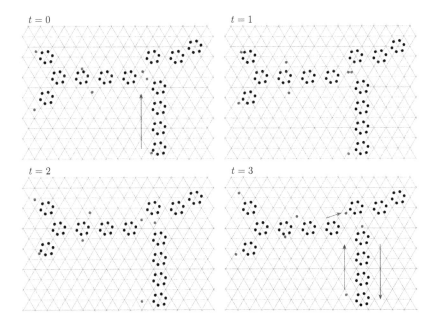

Fig. 7.59   Transferring state signals from a Y-module (left) to a bridge (right). At $t = 0$ a state signal interacts with an input signal coming from the south. Then, the state signal moves to the bridge module.

A flip-flop will be composed of two Y-modules and a bridge module. When the state of the flip-flop changes, five state signals move from one Y-module to another via a bridge. Here, a bridge consists of three blocks.

Figure 7.59 is a process of shifting state signals from a Y-module to a bridge. At $t = 0$ a state signal interacts with an input signal coming from the south. By this, the state signal is directed to the bridge. Giving five consecutive input signals from the south, all the five state signals are transported to the bridge. Shifting state signals from a bridge to a Y-module is performed similarly.

To make a flip-flop, we need two Y-modules, and a bridge as shown in Fig. 7.60. We explain why the bridges has three blocks. As in the case of ETPCA-0157, consider the following process. At first, state signals are on the left Y-module. Giving input signals by the method of Fig. 7.59, the state signals are transferred to the bridge, and then to the right Y-module. The state signals rotate around the right Y-module many times. Then, by giving input signals from the north, the state signals move to the bridge, and then reach the left Y-module.

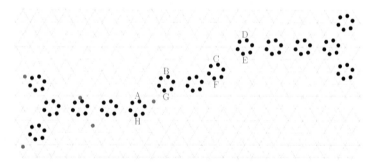

Fig. 7.60   Two Y-modules and a bridge in ETPCA-013z [63].

Let delay(P, Q) denote the minimum time for a signal to reach the point Q from P. Then, we can calculate the following easily.

delay(A, B) = delay(C, D) = delay(E, F) = delay(G, H) = 20
delay(B, C) = delay(F, G) = 20
delay(D, E) = delay(H, A) = 135

Assume a state signal starts the point A at $t = 0$, and returns to A at $t = t_1$ in the above process. Then, t_1 is as follows, where k is a nonnegative integer. Note that the time to go around a Y-module just once is 150 steps.

$$t_1 = 20 \times 6 + 135 \times 2 + 150k$$

We can see that t_1 is divisible by 30, and thus the state signals can be moved to the right again by the control signals of the same phase as those of $t = 0$. It makes it easy to use the module.

If there is no bridge, the returning time t_2 of the state signal is as follows.

$$t_2 = 20 \times 4 + 135 \times 2 + 150k$$

Since t_2 is indivisible by 30, to move the state signals again the phase of the control signals must be shifted. Hence, it becomes more difficult to change the state repeatedly.

Figure 7.61 is a pattern of a *flip-flop* in ETPCA013Z. Assume it is in the state 0, *i.e.*, state signals go around the left Y-module. If five signals are given to the S input port with a right timing, the state signals are shifted to the bridge. Then the input signals are immediately sent to the position between the bridge and the right Y-module. By this, the state signals are further shifted to the right Y-module, and the state becomes 1. Likewise, if five signals are given to R when the state is 1, then it is reset to 0.

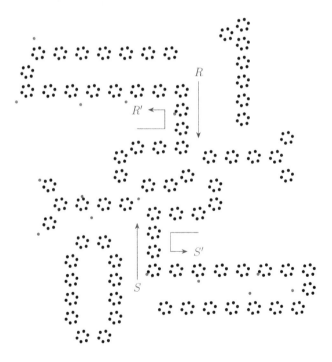

Fig. 7.61   Flip-flop module in ETPCA-013z [63].

To construct RLEM 4-31 by extending a flip-flop module, we need a mechanism that performs branching and merging of signal paths by testing if state signals exist or not on a Y-module. An implementation method of such a mechanism is shown in Fig. 7.62.

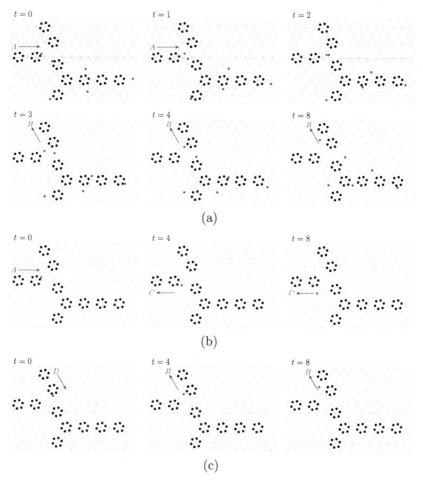

Fig. 7.62   Branching and merging of signal paths in ETPCA-013z. It is performed by testing existence of state signals on a Y-module. (a) Sensing that state signals exist. It is used for both branching and merging. (b) Sensing that no state signal exists. It is for branching. (c) Sensing that no state signal exists. It is for merging.

Giving test signals from the path A, branching is performed. If state signals exist on the Y-module, then the test signals goes to the path B

(Fig. 7.62(a)). On the other hand, if no state signal exists, then the test signals goes to the path C (Fig. 7.62(b)).

The process of Fig. 7.62(a) is also used for merging two signal paths. If signals are given from the path D in Fig. 7.62(c) under the assumption that no state signal exists on the Y-module, then they goes to the path B. By this, a reversible merge of the paths A and D into B is realized.

Figure 7.48 is a pattern that simulates RLEM 4-31 in ETPCA-0157. It has a flip-flop module in its center part. Around it, four sets of branch-and-merge mechanisms are attached. Signals from the input port a, b, c or d first tests whether the state is 0 or 1. Depending on the test result, the state of RLEM 4-31 is changed if it is necessary. Then, the signal paths are merged into four, and finally goes out from the output port w, x, y or z.

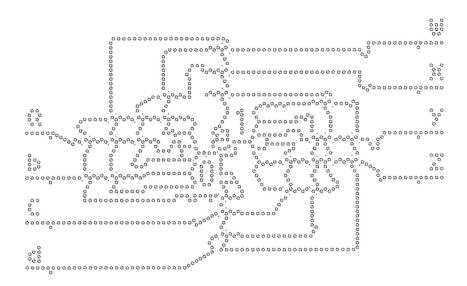

Fig. 7.63   RLEM 4-31 implemented in ETPCA-013z [63].

7.4.3   *Making RTMs by RLEM 4-31 in ETPCA-013z*

Figure 7.64 is a configuration of RTM T_{parity} implemented in ETPCA-013z. It is obtained by replacing each occurrence of RLEM 4-31 in Fig. 4.34 by the pattern of Fig. 7.63 and connecting the patterns appropriately. Since any RTM can be constructed by this method, ETPCA-013z is Turing universal.

Fig. 7.64   RTM T_{parity} in ETPCA-013z simulated on Golly [21].

7.5   Remarks and Notes

A reversible ETPCA was first studied in [30] to investigate the problem how simple a Turing universal reversible CA can be. In [30], it was shown that a Fredkin gate, a universal reversible logic gate, is simulated in ETPCA-0157, which has only four local transition rules. Later, it was also shown that a Fredkin gate is realizable in ETPCA-0347 [33]. By them, we can conclude that ETPCA-0157 and ETPCA-0347 are Turing universal. However, designing full configurations of RTMs is very difficult if we use only logic gates. This is because there is a cumbersome problem where timing of signals must be exactly adjusted at each gate.

This problem was first solved by realizing RLEM 4-31 in ETPCA-0347 [65]. There, full configurations of an RTM were constructed simply and systematically by using the pattern of RLEM 4-31. Furthermore, configurations of the RTM were simulated on Golly. A revised version of the pattern of RLEM 4-31 is shown in Fig. 7.12, which was given in [66]. Figure 7.20 is a configuration of T_{parity} using the revised pattern.

In [63], RLEM 4-31 was implemented in ETPCA-0137, and full configurations of an RTM were designed. In the evolving processes of these configurations, the fourth local transition rule of ETPCA-0137 is not used at all. Therefore, it is actually ETPCA-013z, a partial ETPCA. Figure 7.64 shows an example of an RTM configuration in ETPCA-013z. Thus, we can see that the extremely simple ETPCA having only three local transition rules is Turing universal.

In Sec. 7.3, a pattern that simulates RLEM 4-31 was composed using a technique similar to the one in ETPCA-013z. A configuration of the RTM T_{parity} in ETPCA-0157 is given in Fig. 7.49.

In all the above ETPCAs, RLEM 4-31, rather than an RE, is constructed, though the operations of an RE are understandable more easily than those of RLEM 4-31. The reason is as follows. As discussed in

Sec. 4.3.5.2, RTMs can be constructed out of RLEM 4-31 simply as in the case of using an RE. In addition, since each state of RLEM 4-31 has only one case of a state change as seen in Fig. 7.11, it is easier to implement it in ETPCAs than an RE. In particular, in ETPCAs 0157 and 013z, it is difficult to provide paths for transferring state signals between two Y-modules, if there are two or more state changes in each state.

Since RLEM 4-31 is an intrinsically universal RLEM (Theorem 4.9), any local function of a reversible SPCA or a reversible TPCA is realized by a circuit composed of RLEM 4-31 in a similar manner as in the case of using an RE (see Fig. 6.49). Implementing the circuit in ETPCA-0347, 0157, and 013z, metacell of any reversible SPCA or TPCA can be realized. Therefore, ETPCA-0347, 0157, and 013z are intrinsically universal with respect to the classes of reversible SPCAs and TPCAs.

In Sec. 7.2.2, it is shown that RLEM 2-2 is simulated in ETPCA-034z. It has been shown that RLEM 2-2 is Turing universal (Theorem 4.12), but not intrinsically universal (Theorem 4.10). Therefore, ETPCA-034z is Turing universal, but it is not known whether ETPCA-034z is intrinsically universal. We can see that if ETPCA-034z can simulate either RLEM 2-3 or RLEM 2-4, which are the weakest 2-state RLEMs except RLEM 2-2 (see Fig. 4.22), then it is intrinsically universal. The reason is as follows. Assume RLEM 2-3 (RLEM 2-4, respectively) is realized in ETPCA-034z. Then RLEM 2-4 (RLEM 2-3) is obtained by applying the involution $H^{\mathrm{rev}} \circ H^{\mathrm{refl}}$ to the pattern of RLEM 2-3 (RLEM 2-4) in a similar manner as in Sec. 6.1.3. This is because ETPCA-0347 (and also ETPCA-034z) is T-symmetric under $H^{\mathrm{rev}} \circ H^{\mathrm{refl}}$. Since the set {RLEM 2-3, RLEM 2-4} is intrinsically universal (Lemma 4.7), ETPCA-034z can be intrinsically universal. However, currently it is an open problem.

7.6 Exercises

7.6.1 *Paper-and-pencil exercises*

Exercise 7.1.* Consider ETPCA-0347 and the configuration shown in Fig. 7.65. It contains a fin and a block (see Figs. 7.3 and 7.4). Write the configurations for $0 \leq t \leq 14$. How will it evolve after $t = 14$?

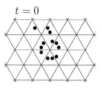

$t = 0$

Fig. 7.65   Configuration consisting of a fin and a block.

Exercise 7.2.* Consider the pushing process of a fin by a glider-6 in ETPCA-0347 given in Fig. 7.10(a). The configuration at $t = 30$ is shown in Fig. 7.66. Write the configurations for $30 \leq t \leq 42$, and verify that the fin is actually pushed by this process.

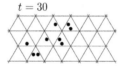

$t = 30$

Fig. 7.66   Pushing a fin by a glider-6.

Exercise 7.3.** Consider ETPCA-0347. Design a pattern that simulates RLEM 2-2 (Fig. 7.67) in it. Note that there may be some difficulty in adjusting the phase of a glider-6 (Table 7.1 and Fig. 7.10 should be referred).

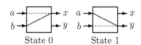

Fig. 7.67   RLEM 2-2.

7.6.2 *Golly exercises*

Exercise 7.4.* Simulate the evolution processes of Exercises 7.1 and 7.2 by Golly.

Exercise 7.5.** Implement the patterns of RLEM 2-2 designed in Exercise 7.3 using Golly, and verify that it works correctly.

Chapter 8

Making Reversible Counter Machines in a Reversible SPCA

In Chaps. 6 and 7, RTMs are constructed in reversible ESPCAs and ETP-CAs. There, a tape cell is realized as a logic circuit composed of logic elements, as in the case of the today's computer memory. Therefore, the resulting configurations of RTMs become infinite (but they are finitely describable).

In this chapter, we investigate a Turing universal reversible PCA with *finite configurations* (see Definition 4.23). So far, it is not known whether such an ESPCA or ETPCA exists. We show here a particular 81-state reversible SPCA P_3 with finite configurations that can simulate any reversible counter machine (RCM) [29]. In a cell of P_3 each of four parts has three states. By this it is named P_3. Since the class of RCMs has been shown to be Turing universal (Theorem 4.5), Turing universality of P_3 follows.

As we shall see in Sec. 8.1, P_3 has a property similar to a conservative ESPCA (Definition 2.9). It means that, if an initial configuration is finite, then the total number of particles in a configuration is bounded throughout its evolution process. In P_3, a contents of a counter (*i.e.*, a nonnegative integer) will be kept by a distance between the origin and a position marker. Increment and decrement of the counter is performed by shifting the position marker. By this, a counter is implemented using a fixed number of particles.

8.1 Reversible Rotation-Symmetric 81-State SPCA P_3

An SPCA having the state set $\{0, 1, 2\}$ in each part of a cell is defined by
$$P = (\mathbb{Z}^2, (T, R, B, L), ((0, -1), (-1, 0), (0, 1), (1, 0)), f, (0, 0, 0, 0))$$
where $T = R = B = L = \{0, 1, 2\}$ (see Definition 2.1). Hence, P is an 81-state SPCA. Here, we impose the restriction of rotation-symmetry

(Definition 2.7) to P. This class of 81-state rotation-symmetric SPCAs is considered to be an extension of the class of ESPCAs.

We introduce a particular 81-state rotation-symmetric SPCA P_3 having the local function f_{P_3} shown in Fig. 8.1. In this figure, the states 0, 1, and 2 are represented by a blank, a white particle, and a black particle, respectively. Since P_3 is rotation-symmetric, its local function is completely described by the 24 local transition rules given in Fig. 8.1. If a cell is in the state $(t, r, b, l) \in \{0, 1, 2\}^4$, it can be represented by an integer $3^3 t + 3^2 r + 3^1 b + 3^0 l$, a *state number*. The number in the right-hand side of each local transition rule in Fig. 8.1 shows the state number. Thus, f_{P_3} is described by the following 24-tuple of state numbers.

$$(0, 1, 36, 30, 37, 40, 6, 56, 60, 62, 80, 45, 63, 19, 34, 42, 43, 75, 47, 16, 44, 50, 73, 71)$$

Any other 81-state rotation-symmetric SPCA is also described by a 24-tuple of state numbers. However, since such a description is very long, we use a short name like P_3.

From Fig. 8.1, we can verify that no pair of local transition rules, including rotated ones, have the same right-hand side. Therefore, f_{P_3} is injective, and hence P_3 is a reversible SPCA (Theorem 2.1).

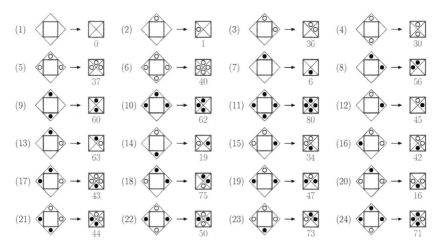

Fig. 8.1 Local transition rules that define the local function f_{P_3} of the 81-state rotation-symmetric reversible SPCA P_3 [29]. The states 0, 1, and 2 are represented by a blank, a white particle ○, and a black particle ●, respectively.

We can see that each local transition rule conserves the total number. For example, consider the rule (17). Though the number of particles is not conserved in (17), the total of the numbers of the four parts in the left-hand side (*i.e.*, $2+2+1$) is equal to that in the right-hand side (*i.e.*, $1+1+2+1$). A PCA that satisfies this condition is called a *number-conserving PCA*. It is an extension of the notion of conservative PCA (Definition 2.9).

We can also see that in Fig. 8.1, 12 local transition rules (1), (3), (4), (5), (6), (9), (11), (12), (13), (19), (21) and (22) have a simple property, where each incoming particle in the left-hand side simply turns backward in the right-hand side. Therefore, they are described by the rule scheme (25) of Fig. 8.2. Thus, we have a simplified description of f_{P_3} in Fig. 8.2.

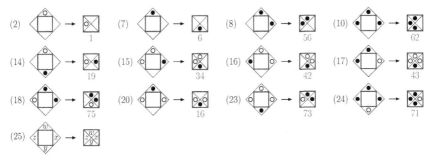

Fig. 8.2   Simplified description of the local function f_{P_3} of P_3. The rule scheme (25) represents 12 local transition rules not specified by the other rules, where $w, x, y, z \in \{\text{blank}, \circ, \bullet\}$.

Since the number of local transition rules of an 81-state rotation-symmetric SPCA is much larger than that of an ESPCA or an ETPCA, the local function of the former can be designed in a somewhat "intended" manner. Actually, in P_3, its local function f_{P_3} is defined so that an RE, a position marker, and others are implemented easily, as we shall see below.

On the other hand, in an ESPCA or an ETPCA, we can only "select" interesting local functions, and try to find useful phenomena in the cellular space for composing RTMs. Designing a local function of an ESPCA or an ETPCA in an intended way is very difficult, since interactions of small patterns often show complex behavior that lasts a long time (for example, a shifting process of a position marker in Fig. 6.20 takes more than 2000 steps), and thus they are unpredictable for a human observer.

Of course, in an 81-state rotation-symmetric SPCA, complex phenomena often appear. However, P_3 is an artificially designed SPCA, and thus basic gadgets that will be used to compose RCMs do not show complex behavior. This is the main difference from the cases of ESPCA and ETPCA.

8.2   Useful Gadgets in the SPCA P_3

In P_3, a single black particle acts as a space-moving pattern of period 1 as shown in Fig. 8.3. It is called a *glider-1* in P_3. To construct RCMs, we use a glider-1 as a *signal*.

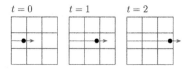

Fig. 8.3   Glider-1 that is used as a signal in SPCA P_3 [29].

We introduce five kinds of useful gadgets for composing RCMs in P_3. The first one is an *LR-turn gadget* composed of eight white particles shown in Fig. 8.4. Colliding a signal with an LR-turn gadget as in Fig. 8.4(a), a left-turn of a signal is realized. A right-turn of a signal is also possible as in Fig. 8.4(b).

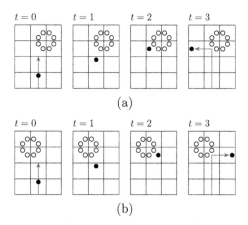

Fig. 8.4   LR-turn gadget in SPCA P_3 [29]. (a) Left-turn, and (b) right-turn of a signal.

The second is an *R-turn gadget*. A signal can make a right-turn by the process shown in Fig. 8.5. This gadget will be used as an interface between unidirectional signal paths and a bidirectional signal path as in Fig. 8.6. As it will be seen below, an RE gadget and pushing/pulling a position marker use a bidirectional signal path on which a signal can move in both of the directions. Sending or receiving a signal to/from a bidirectional signal path, an R-turn gadget will be used.

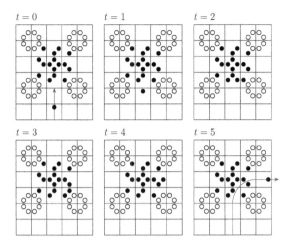

Fig. 8.5   R-turn gadget in SPCA P_3 [29].

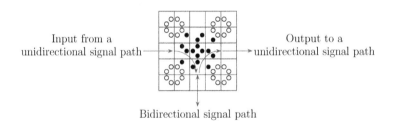

Input from a
unidirectional signal path

Output to a
unidirectional signal path

Bidirectional signal path

Fig. 8.6   Interface between unidirectional signal paths and a bidirectional signal path composed of an R-turn gadget in SPCA P_3.

The third is a *B-turn gadget*. By this, a signal makes a backward-turn as shown in Fig. 8.7. Although a backward-turn can be realized by LR-turn and R-turn gadgets, it often reduces the size of a larger module composed of many gadgets.

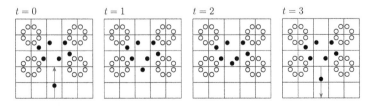

Fig. 8.7   B-turn gadget in SPCA P_3 [29].

The fourth is an *RE gadget*. It simulates a rotary element (RE). The pattern of Fig. 8.8(a) (Fig. 8.8(b), respectively) stands for the state V (state H) of an RE. It should be noted that the input/output signal lines of the original RE are unidirectional, while those of the RE gadget are bidirectional. Therefore, to use the RE gadget, an interface between unidirectional signal lines and a bidirectional signal line shown in Fig. 8.6 is required.

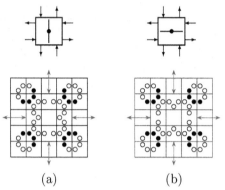

(a)   (b)

Fig. 8.8   RE gadget in SPCA P_3 [29]. (a) State V, and (b) state H. It has four bidirectional input/output signal lines.

Figure 8.9 shows how the pattern of an RE gadget evolves when it is in the state V and an input signal comes from the south.

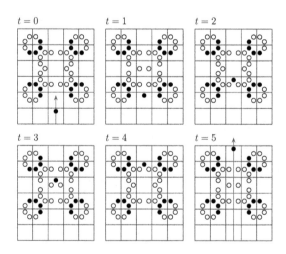

Fig. 8.9   Operation of an RE gadget in SPCA P_3: the parallel case [29].

Figure 8.10 shows the case where the state is H and an input signal comes from the south. In this case, the state changes to V, and the signal goes out from the east port.

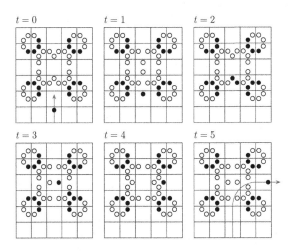

Fig. 8.10   Operation of an RE gadget in SPCA P_3: the orthogonal case [29].

The fifth is a *position marker*. It consists of one white particle, and thus it rotates with a period of 4 as shown in Fig. 8.11.

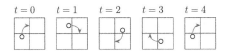

Fig. 8.11   Position marker in SPCA P_3 [29].

Colliding a signal with a position marker at a right timing, the marker position is pushed by one cell as shown in Fig. 8.12.

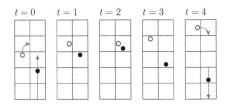

Fig. 8.12   Pushing a position marker in SPCA P_3 [29].

On the other hand, colliding a signal as in Fig. 8.13, the position of the marker is pulled by one cell.

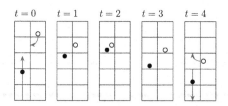

Fig. 8.13   Pulling a position marker in SPCA P_3 [29].

By above, we can see that a position marker can be used for realizing a counter of an RCM. Namely, an integer stored in a counter is kept by the distance between the marker and a properly defined origin. Increment and decrement operations are performed by pushing and pulling the position marker. Note that how to test whether the integer is 0 or positive is explained in the next section.

8.3   Designing a Counter Module in P_3

We construct a counter module that simulates a counter of an RCM. Though it is composed of a fixed number of particles in P_3, it can store an arbitrarily large nonnegative integer.

Here, RCMs in the program form defined in Sec. 4.2.3 is constructed. An RCM in the program form has the following four kinds of instructions for the i-th counter, where b_0, b_1, m_0 and m_1 are addresses of instructions.

I_i	Increment the i-th counter
D_i	Decrement the i-th counter
$B_i(b_0, b_1)$	Test and branch on the contents of the i-th counter, *i.e.*, if the i-th counter is 0, then go to b_0, else go to b_1
$M_i(m_0, m_1)$	Merge on the contents of the i-th counter, *i.e.*, if the i-th counter is 0, then merge from m_0, else from m_1

The counter module is designed so that the above instructions can be executed easily by a finite control module given in the next section. Note that, besides the above four, there is a halt instruction H. However, when an RCM halts, a pattern that simulates the RCM will emit a halt signal outside. Therefore, the halt instruction is not managed in a counter module.

Figure 8.14 is a *counter module* implemented in the reversible 81-state SPCA P_3. It is composed of two position markers (labelled by m_1 and m_2), twelve RE gadgets (labelled by **a**, **b**, ..., **k** and **l**), six R-turn gadgets, two B-turn gadgets, and many LR-turn gadgets. To make an RCM(k), k copies of a counter module are used. Since it has position markers of period 4, an input signal must be given at $t \equiv 0 \pmod 4$.

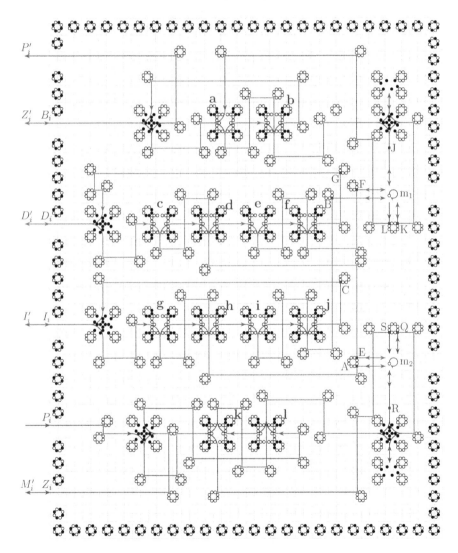

Fig. 8.14　Counter module for the i-th counter realized in the 81-state SPCA P_3 [29].

The reason why two position markers are used is for keeping a sufficient number of access paths to execute $B_i(b_0, b_1)$ and $M_i(m_0, m_1)$ instructions. In Fig. 8.14, the two markers are placed at the origin positions. If the instruction I_i (D_i, respectively) is executed, then the markers are shifted to the right (left) by two cells. This is for adjusting the phase of markers. Thus, if the markers at the positions of $2n$ cells east from the origin, we interpret that the counter contains the integer n.

The counter module has the following ten input and output ports for a signal. Note that among them four pairs of input ports and output ports share bidirectional signal paths. Hence, a counter module has six input/output signal lines in total.

I_i: Input port for the I_i instruction
I_i':   Output port for telling completion of the I_i instruction
D_i: Input port for the D_i instruction
D_i':   Output port for telling completion of the D_i instruction
B_i: Input port for the $B_i(b_0, b_1)$ instruction
Z_i':   Output port for telling the i-th counter is 0
P_i':   Output port for telling the i-th counter is positive
M_i':   Output port for the $M_i(m_0, m_1)$ instruction
Z_i: Input port for $M_i(m_0, m_1)$ when the i-th counter is 0
P_i: Input port for $M_i(m_0, m_1)$ when the i-th counter is positive

We now explain how the four instructions are executed by the counter module. They will be used as "subroutines" called from a finite control. We assume, initially, all the eight REs are in the state V.

(1) Execution of the instruction I_i:
If we give a signal from the input port I_i, we obtain the following sequence that shows how the signal travels.

$(I_i, \mathbf{g}_H, \mathbf{h}_V, A, m_2^+, A, \mathbf{h}_V, \mathbf{g}_V, \mathbf{h}_H, A, m_2^+, A, \mathbf{h}_V,$
    $\mathbf{i}_H, \mathbf{j}_V, B, m_1^+, B, \mathbf{j}_V, \mathbf{i}_V, \mathbf{j}_H, B, m_1^+, B, \mathbf{j}_V, C, I_i')$

It also describes operations performed by a signal as explained below.

If a signal is given to the port I_i, then it goes to the RE labelled by $\mathbf{g}$, and changes its state to H. Here, $\mathbf{g}_H$ means that after the signal accesses the RE $\mathbf{g}$, it becomes the state H. Next, the signal goes to the RE $\mathbf{h}$ leaving its state V unchanged, and reaches the point A. From A, the signal pushes the position marker m_2 rightward, where m_2^+ means that the marker m_2 is pushed. The signal goes back to A, and passes through the RE $\mathbf{h}$ leaving its state V unchanged. Next, it goes to the RE $\mathbf{g}$, and restores its state to V. The signal accesses the RE $\mathbf{h}$, and changes its state to H. Then, from the point A, the signal pushes the position marker m_2 again. The signal goes back to A, and restores the state of $\mathbf{h}$ to V. By above, the marker m_2 is shifted rightward by two cells. Note that the two REs labelled by $\mathbf{g}$ and $\mathbf{h}$ are used to execute the pushing process of the marker m_2 twice.

The latter half of the sequence starting from $\mathbf{i}_H$ is for pushing the marker m_1 rightward twice. It is performed by using the REs $\mathbf{i}$ and $\mathbf{j}$ in a similar manner as in the case of the former half of the sequence. At the end of this sequence, the signal reaches the point C, and finally goes out from the port I_i'. By above, execution of the instruction I_i is completed.

(2) Execution of the instruction D_i:

If we give a signal from the input port D_i, we have the following sequence.

$(D_i, \mathbf{c}_H, \mathbf{d}_V, E, m_2^-, E, \mathbf{d}_V, \mathbf{c}_V, \mathbf{d}_H, E, m_2^-, E, \mathbf{d}_V,$

$\quad \mathbf{e}_H, \mathbf{f}_V, F, m_1^-, F, \mathbf{f}_V, \mathbf{e}_V, \mathbf{f}_H, F, m_1^-, F, \mathbf{f}_V, G, D_i')$

The sequence is similar to the case of I_i except that the position markers are pulled by m_1^- and m_2^-. Here, the RE $\mathbf{c}$ and $\mathbf{d}$ are used for pulling m_2 twice, while the RE $\mathbf{e}$ and $\mathbf{f}$ are for pulling m_1 twice.

(3) Execution of the instruction $B_i(b_0, b_1)$:

This instruction is performed by testing whether the position marker m_1 is at the origin position or not.

When the counter keeps 0, we have the following sequence by giving a signal from the input port B_i.

$(B_i, \mathbf{a}_H, \mathbf{b}_V, J, m_1^+, J, K, m_1^+, K, \mathbf{b}_V, \mathbf{a}_V, \mathbf{b}_H, J, m_1^+, J, K, m_1^+, K, \mathbf{b}_V, Z_i')$

The signal first changes the state of the RE $\mathbf{a}$ to H, then passes through the RE $\mathbf{b}$ without changing its state V, and reach the point J. Since the counter keeps 0, the signal from J pushes the marker m_1 downward, and goes back to J. From the point K, the signal pushes the marker m_1 upward, and then goes back to K. By this, m_1 returns to the original position. However, the phase of the marker m_1 is shifted by 2. Therefore, the above process should be repeated once again to restore the phase. The latter half of the sequence does so. The two REs $\mathbf{a}$ and $\mathbf{b}$ are used to perform the above process twice. Finally, the signal goes out from the output port Z_i'.

When the counter keeps a positive integer, we have the following sequence by giving a signal from the input port B_i.

$(B_i, \mathbf{a}_H, \mathbf{b}_V, J, L, \mathbf{a}_V, P_i')$

The signal first changes the state of the RE $\mathbf{a}$ to H, then passes through the RE $\mathbf{b}$, and reach the point J as the first case. This time, since the counter keeps a positive integer, the signal goes to the point L. Then the signal restores the state of the RE $\mathbf{a}$ to V, and goes out from the output port P_i'.

By above, branching the signal path depending on the contents of the counter is correctly performed. Jumping to the address b_0 or b_1 in the instruction $B_i(b_0, b_1)$ will be performed by a finite control of an RCM.

(4) Execution of the instruction $M_i(m_0, m_1)$:
This instruction is performed by a sequence of operations that is similar to the inverse sequence of operations for the instruction $B_i(b_0, b_1)$. It refers the position marker m_2.

When the counter keeps 0, we have the following sequence by giving a signal from the input port Z_i.

$(Z_i, l_H, Q, m_2^+, Q, R, m_2^+, R, l_V, k_H, l_V, Q, m_2^+, Q, R, m_2^+, R, l_V, k_V, M_i')$

The signal first goes to the point Q via the RE l, and pushes the marker m_2 downward. Then it goes to the point R, and pushes m_2 upward to restore its position. This process is repeated once again using the REs k and l. By this, the phase of m_2 is adjusted. Finally the signal goes out from M_i.

When the counter keeps a positive integer, we have the following sequence by giving a signal from the input port P_i.

$(P_i, k_H, S, R, l_V, k_V, M_i')$

The signal first goes to the point S via the RE k. Since the counter keeps a positive integer, the signal goes to R, then RE l and k, and finally goes out from M_i'.

By above, reversible merge of the two signal paths into one by referring the contents of the counter is performed.

8.4 Designing a Finite Control Module in P_3

We introduce four kinds of *instruction modules*: an I_i-module, a D_i-module, a B_i-module, and an M_i-module for the i-th counter (Fig. 8.15). They are assembled to compose a finite control module of an RCM in the program form. Each of them has the same names of input/output ports as a counter module. For example, the I_i-module has an input port I_i and an output port I_i', which play the same roles as I_i and I_i' in the counter module. However, the latter ones are used as a calling port and a return port of a subroutine. Thus, these ports will be accessed by many copies of the I_i-module. The RE gadget contained in the I_i-module controls the subroutine call (see the call mechanism given in Fig. 4.26). Note that, besides the four, an L-module (Fig. 8.15) is also used, which is placed at the left end of the finite control module.

Below, we explain how the instruction modules work. We assume, initially, all the REs contained in the modules are in the state H.

(1) The I_i-module acts as follows. When a signal is given to the input port I_i, it first changes the state of the RE to V, and then it calls the subroutine I_i of a counter module. By this, the counter is incremented

by 1. Note that, even if there are other occurrences of I_i-modules, the signal does not affect them, since REs in them are in the state H. If a signal goes back from the port I_i' of the counter module, the RE is restored to the state H, and it goes out from the output port I_i'.

(2) The D_i-module acts similarly as the I_i-module, except that it calls the subroutine D_i of a counter module.

(3) The B_i-module works as follows. When a signal is given to B_i, it first sets the RE to the state V, and calls the subroutine B_i of the counter module. If the signal comes back from Z_i' of the counter module, it restores the RE to the state H, and then goes out from the output port Z_i'. If the signal comes back from P_i' of the counter module, it is reflected at the L-module, and then reaches to the RE of the module. It restores the RE to the state H, and then goes out from the output port P_i'.

(3) The M_i-module performs the inverse operations of the B_i-module. Namely, two signal paths from Z_i and P_i are merged into the path to M_i'.

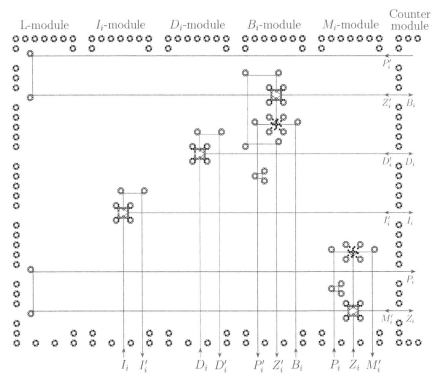

Fig. 8.15   Four instruction modules for the i-th counter, and an L-module in P_3.

If an RCM M in the program form is given, we can compose a pattern of its finite control by the following method. First, prepare instruction modules for all the instructions contained in the M's program P (except H instructions), and place them in the same order as in P. Also, put an L-module at the left end. Then, connect the instruction modules by signal paths based on the order of the instructions, and the source and destination addresses in M_i and B_i instructions. Finally, add a signal path from a designated input port "Start" to the first instruction module, and a path from the instruction that is followed by an H instruction to a "Halt" output port (if there are many H instructions in P, many Halt ports should be prepared). We explain the method by an example.

Example 8.1. Consider an RCM(2) $\hat{M}_{\text{twice}}$ in Example 4.9. Figure 8.16 shows its well-formed program (WFP) $\hat{P}_{\text{twice}}$. For a given nonnegative integer n, $\hat{M}_{\text{twice}}$ computes the function $g(n) = 2n$, *i.e.*, it performs the computation $(0, (n, 0)) \mathbin{\vert_{\hat{M}_{\text{twice}}}^{*}} (9, (0, 2n))$.

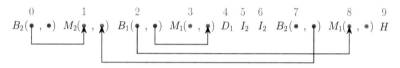

Fig. 8.16   WFP $\hat{P}_{\text{twice}}$ in Example 4.9.

We can implement the WFP $\hat{P}_{\text{twice}}$ in the SPCA P_3 using instruction modules, but we first simplify $\hat{P}_{\text{twice}}$ by an *ad hoc* method, and then implement it in P_3. Each of the two M_1 instructions at the addresses 3 and 8 has only one source address, and there is no need to merge the execution paths at the M_1 instruction. Thus, we remove the two M_1 instructions, and change the destination addresses pointing to them. By this, we have a simplified program $\hat{P}_{\text{twice}}^{\dagger}$ (Fig. 8.17). Let $\hat{M}_{\text{twice}}^{\dagger}$ be an RCM(2) having $\hat{P}_{\text{twice}}^{\dagger}$. Although it is not a WFP (Definition 4.12), it is executable without ambiguity, and hence we implement it in P_3. However, we do not change Definition 4.12. This is for keeping conciseness of the definition, and keeping symmetry between B_i and M_i instructions.

Fig. 8.17   Program $\hat{P}_{\text{twice}}^{\dagger}$, which is not a WFP, obtained by simplifying $\hat{P}_{\text{twice}}$.

Figure 8.18 is a finite control module of $\hat{M}^{\dagger}_{\text{twice}}$ implemented in the SPCA P_3. Instruction modules B_2, M_2, B_1, D_1, I_2, I_2, and B_2 are placed according to the order in $\hat{P}^{\dagger}_{\text{twice}}$. Modules for the first counter are in the upper part of the pattern, while those for the second counter is in the middle part. In the lower part, transitions among the instructions are realized, and the Start and Halt ports are placed by laying signal paths.

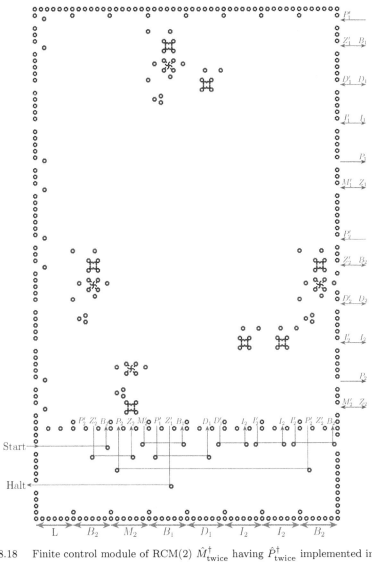

Fig. 8.18   Finite control module of RCM(2) $\hat{M}^{\dagger}_{\text{twice}}$ having $\hat{P}^{\dagger}_{\text{twice}}$ implemented in P_3.

For example, the Start port is connected to the B_2 port of the first B_2 module, and its output port Z_2' is connected to the input port Z_2 of the M_2 module, and so on. The output port of the B_1 module is connected to the Halt port. In this way, the flow of the program $\hat{P}_{\text{twice}}^{\dagger}$ is realized.

8.5   Implementing RCMs as Finite Configurations of P_3

Attaching two counter modules to the finite control module given in Fig. 8.18, we obtain a finite configuration that simulates the RCM(2) $\hat{M}_{\text{twice}}^{\dagger}$. Figure 8.19 shows the finite configuration realized in the reversible 81-state SPCA P_3 [21].

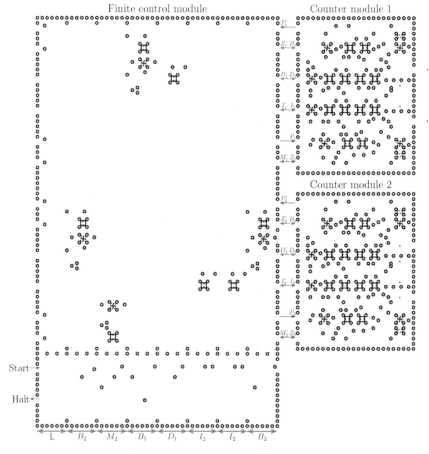

Fig. 8.19   Full configuration of RCM(2) $\hat{M}_{\text{twice}}^{\dagger}$ implemented in P_3.

By the above method, we can compose a configuration of any RCM with an arbitrary number of counters. Thus, the SPCA P_3 *with finite configurations* is Turing universal.

Example 8.2. Consider the RCM(3) $M_{\exp}$ in Example 4.5. It computes the function $g(n) = 2^n$, *i.e.*, it performs the following computation: $(q_0, (n, 0, 0)) \vdash^*_{M_{\exp}} (q_f, (0, 2^n, 0))$. By the method given in Theorem 4.7 and in Example 4.9, we have an RCM(3) $M'_{\exp}$ with a WFP $P_{\exp}$ (Fig. 8.20). By removing M_i instructions at the addresses 4, 8, 14, 18 and 23 of $P_{\exp}$ as in Example 8.1, we have a simplified program $P^{\dagger}_{\exp}$. Figure 8.21 shows a configuration that simulates $M^{\dagger}_{\exp}$ with $P^{\dagger}_{\exp}$ [21].

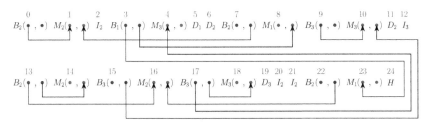

Fig. 8.20   WFP $P_{\exp}$ that computes the function $g(n) = 2^n$.

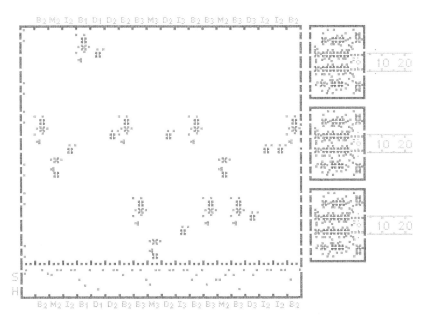

Fig. 8.21   RCM(3) $M^{\dagger}_{\exp}$ with $P^{\dagger}_{\exp}$ realized in the SPCA P_3 simulated on Golly.

8.6   Composing Inverse RCM in P_3

In this section, by using the time-reversal symmetry (T-symmetry) of P_3 shown below, we construct an *inverse RCM* that *un*-computes the function computed by a given RCM. Similar to the cases of ESPCAs and ETPCAs, some reversible 81-state SPCAs are T-symmetric, *i.e.*, their local functions for the negative time direction are essentially the same as the ones for the positive time direction. We give here two involutions H^{rev} and H^{refl} that are similarly defined as in ESPCAs (Sec. 3.2), and show that P_3 is T-symmetric under $H^{\mathrm{rev}} \circ H^{\mathrm{refl}}$.

For a local function $f : \{0, 1, 2\}^4 \to \{0, 1, 2\}^4$ of an 81-state SPCA, the global function induced by f is denoted by F_f. Note that the global function is defined in Definition 2.5.

Let $f_{\mathrm{rev}} : \{0, 1, 2\}^4 \to \{0, 1, 2\}^4$ be a local function that satisfy the following (see Fig. 8.22).

$$\forall (t, r, b, l) \in \{0, 1, 2\}^4 \ (f_{\mathrm{rev}}(t, r, b, l) = (b, l, t, r))$$

It makes every particle coming to a cell go back to the opposite direction. We define the involution H^{rev} as follows.

$$H^{\mathrm{rev}} = F_{f_{\mathrm{rev}}}$$

It is called the *involution of motion reversal*. By H^{rev}, moving directions of all particles in a configuration are reversed.

     for all $(t, r, b, l) \in \{0, 1, 2\}^4$

Fig. 8.22   Local function f_{rev} that reverses moving directions of particles.

The following lemma can be proved in a similar manner as in Lemma 3.1, and thus its proof is omitted.

Lemma 8.1. *Let $f : \{0, 1, 2\}^4 \to \{0, 1, 2\}^4$ be a local function of an 81-state reversible SPCA. Then, the following holds.*

$$F_f^{-1} = H^{\mathrm{rev}} \circ F_{f^{-1}} \circ H^{\mathrm{rev}}$$

By the above lemma, we have the following.

Theorem 8.1. *Let $f : \{0, 1, 2\}^4 \to \{0, 1, 2\}^4$ be a local function of an 81-state reversible SPCA. If $f^{-1} = f$, then the following holds.*

$$F_f^{-1} = H^{\mathrm{rev}} \circ F_f \circ H^{\mathrm{rev}}$$

We define the notion of strict T-symmetry as follows.

Definition 8.1. Let P be a reversible SPCA whose local function is f. If $F_f^{-1} = H^{\text{rev}} \circ F_f \circ H^{\text{rev}}$, then P is called *T-symmetric under H^{rev}*, or *strictly T-symmetric*.

The next definition is analogous to Definition 3.1 for ESPCA.

Definition 8.2. Let $f : \{0,1,2\}^4 \to \{0,1,2\}^4$ be a local function of an 81-state SPCA P. Define $f^{\text{r}} : \{0,1,2\}^4 \to \{0,1,2\}^4$ as follows.
$$\forall (t,r,b,l), (t',r',b',l') \in \{0,1\}^4 :$$
$$f(t,r,b,l) = (t',r',b',l') \Leftrightarrow f^{\text{r}}(t,l,b,r) = (t',l',b',r')$$
The SPCA having the local function f^{r} is denoted by P^{r}, and it is called the *dual SPCA of P under reflection*.

Example 8.3. Consider the SPCA P_3 with the local function f_{P_3} shown in Fig. 8.1. By Definition 8.2, we have the local function $f_{P_3}^{\text{r}}$ as in Fig. 8.23. Thus, the SPCA P_3^{r} with $f_{P_3}^{\text{r}}$ is a dual SPCA of P_3 under reflection.

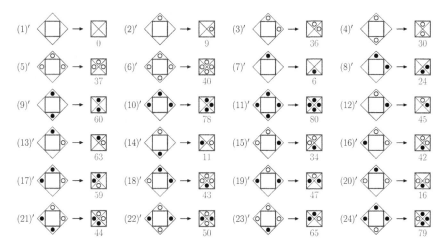

Fig. 8.23   Local function $f_{P_3}^{\text{r}}$ of P_3^{r} that is a dual SPCA of P_3 under reflection.

We give the involution H^{refl} similarly to the case of ESPCA. First, define a function $\text{refl}_4 : \{0,1,2\}^4 \to \{0,1,2\}^4$ as follows: $\text{refl}_4(t,r,b,l) = (t,l,b,r)$ for all $(t,r,b,l) \in \{0,1,2\}^4$. Next, define H^{refl} as follows. For all $\alpha : \mathbb{Z}^2 \to \{0,1,2\}^4$ and $(x_0,y_0) \in \mathbb{Z}^2$:
$$H^{\text{refl}}(\alpha)(x_0,y_0) = \text{refl}_4(\alpha(-x_0,y_0))$$

It is called the *involution of reflection,* and gives the mirror image of a configuration with respect to the y-axis.

The following lemma can be proved in a similar manner as in Lemma 3.2, and thus its proof is omitted.

Lemma 8.2. *Let* $f : \{0,1,2\}^4 \to \{0,1,2\}^4$ *be a local function of any 81-state SPCA. Then, the following holds.*

$$F_{f^r} = H^{\text{refl}} \circ F_f \circ H^{\text{refl}}$$

By Lemmas 8.1 and 8.2, we have the following.

Theorem 8.2. *Let* $f : \{0,1,2\}^4 \to \{0,1,2\}^4$ *be a local function of an 81-state reversible SPCA. If* $f^{-1} = f^r$, *then the following holds.*

$$\begin{aligned} F_f^{-1} &= H^{\text{rev}} \circ F_{f^r} \circ H^{\text{rev}} \\ &= H^{\text{rev}} \circ H^{\text{refl}} \circ F_f \circ H^{\text{refl}} \circ H^{\text{rev}} \end{aligned}$$

Definition 8.3. Let P be a reversible SPCA whose local function is f. If $F_f^{-1} = H^{\text{rev}} \circ H^{\text{refl}} \circ F_f \circ H^{\text{refl}} \circ H^{\text{rev}}$, then P is called *T-symmetric under the involution* $H^{\text{rev}} \circ H^{\text{refl}}$

Example 8.4. Consider the 81-state reversible SPCA P_3 with the local function f_{P_3}. We can verify that $f_{P_3}^{-1} = f_{P_3}^r$. For example,

$$f_{P_3}(0,2,2,1) = (1,1,2,1)$$

holds (see (17) in Fig. 8.1). Therefore,

$$f_{P_3}^{-1}(1,1,2,1) = (0,2,2,1)$$

On the other hand,

$$f_{P_3}^r(2,1,1,1) = (2,1,0,2)$$

holds (see (23)′ in Fig. 8.23). Since, $f_{P_3}^r$ is rotation-symmetric,

$$f_{P_3}^r(1,1,2,1) = (0,2,2,1)$$

Thus, $f_{P_3}^{-1}(1,1,2,1) = f_{P_3}^r(1,1,2,1)$. Other cases are similar.

By Theorem 8.2, we have the following relation.

$$F_{f_{P_3}}^{-1} = H^{\text{rev}} \circ H^{\text{refl}} \circ F_{f_{P_3}} \circ H^{\text{refl}} \circ H^{\text{rev}}$$

Therefore, the SPCA P_3 is T-symmetric under $H^{\text{rev}} \circ H^{\text{refl}}$.

The next lemma is easily proved as in Lemma 3.5.

Lemma 8.3. *Let $f : \{0,1,2\}^4 \to \{0,1,2\}^4$ be a local function of an 81-state reversible SPCA P. Assume P is T-symmetric under the involution H, i.e., $F_f^{-1} = H \circ F_f \circ H$. Then the following holds for any $n \in \{1,2,\ldots\}$.*

$$(F_f^{-1})^n = H \circ (F_f)^n \circ H$$

Using this lemma, we can obtain an *inverse RCM* from a given RCM implemented in SPCA P_3. We explain it by an example.

Example 8.5. Consider the configuration of Fig. 8.21 in which the RCM(3) $M_{\exp}^\dagger$ is implemented. We denote the configuration at $t = 0$ by $\alpha(0)$. If $M_{\exp}^\dagger$ in $\alpha(0)$ keeps a triplet of integers $(m,0,0)$ in its counters, it will eventually enter a halting state at time, say, t_m with a triplet $(0,2^m,0)$. For example, $t_7 = 1{,}496{,}589$. Let $\alpha(t_m)$ denote the configuration at $t = t_m$. If we want have an inverse RCM(3) $(M_{\exp}^\dagger)^{-1}$ that undoes the operations of $M_{\exp}^\dagger$ in P_3, it is obtained by applying $H^{\mathrm{rev}} \circ H^{\mathrm{refl}}$ to $\alpha(t_m)$. This is because P_3 is T-symmetric under $H^{\mathrm{rev}} \circ H^{\mathrm{refl}}$ as explained in Example 8.4. Figure 8.24 shows the configuration $H^{\mathrm{rev}} \circ H^{\mathrm{refl}}(\alpha(t_m))$ [21].

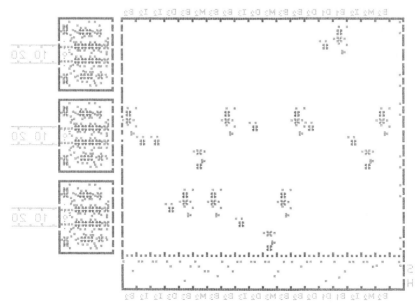

Fig. 8.24　RCM(3) $(M_{\exp}^\dagger)^{-1}$ realized in the SPCA P_3 simulated on Golly. It is obtained by applying the involution $H^{\mathrm{rev}} \circ H^{\mathrm{refl}}$ to the configuration of $M_{\exp}^\dagger$ in Fig. 8.21.

Let $\hat{H} = H^{\mathrm{rev}} \circ H^{\mathrm{refl}}$. Going back to the past from $u(t_m)$ by n steps is simulated by evolving $\hat{H}(\alpha(t_m))$, which is the configuration of Fig. 8.24, by n steps in P_3. In fact, by Lemma 8.3, the following relation holds.

$$(F_{P_3}^{-1})^n(\alpha(t_m)) = \hat{H} \circ (F_{P_3})^n(\hat{H}(\alpha(t_m)))$$

In particular, if $n = t_m$,

$$\alpha(0) = (F_{P_3}^{-1})^{t_m}(\alpha(t_m)) = \hat{H} \circ (F_{P_3})^{t_m}(\hat{H}(\alpha(t_m)))$$

holds, and thus it goes back to the configuration corresponding to the initial configuration $\alpha(0)$. This relation is depicted as a diagram in Fig. 8.25. In this way, an inverse RCM is obtained by a mechanical procedure of applying the involution $\hat{H}$.

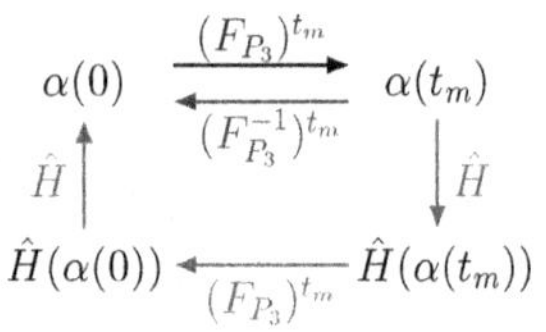

$$
\begin{array}{ccc}
\alpha(0) & \xrightleftharpoons[(F_{P_3}^{-1})^{t_m}]{(F_{P_3})^{t_m}} & \alpha(t_m) \\[2mm]
\hat{H} \uparrow & & \downarrow \hat{H} \\[2mm]
\hat{H}(\alpha(0)) & \xleftarrow{(F_{P_3})^{t_m}} & \hat{H}(\alpha(t_m))
\end{array}
$$

Fig. 8.25   Diagram that shows the relation between the computing process of $M_{\mathrm{exp}}^{\dagger}$ (from $\alpha(0)$ to $\alpha(t_m)$) and the *un*-computing process of $(M_{\mathrm{exp}}^{\dagger})^{-1}$ (from $\hat{H}(\alpha(t_m))$ to $\hat{H}(\alpha(0))$) in the reversible SPCA P_3.

8.7   Remarks and Notes

The problem of finding a Turing universal reversible CA with *finite configurations* was first studied in [67]. There, a 256-state reversible and number-conserving SPCA P_4 is proposed, and it is shown that any RCM can be simulated in a finite configuration of it. It is a rotation-symmetric SPCA in which each part has four states, and thus it is named P_4. Later, this result was improved in [29]. This is the Turing universal 81-state reversible and number-conserving PCA P_3 explained in this chapter. The counter module given in Fig. 8.14 is the same as in [29]. The design of a finite control module (Fig. 8.18) is revised in this chapter so that any RCM in the program form is directly realized.

In a number-conserving PCA, every state of four parts of a cell is represented by a nonnegative integer, where the number 0 is assigned to a quiescent state. In addition, the total of the number is conserved throughout an evolution process, if the configuration is finite. Therefore, if we try to create a memory having unbounded capacity in a finite configuration, the content of the memory is expressed only by a distance between some object pattern and a designated origin, like a counter as shown in Fig. 8.14.

On the other hand, if a CA is not conservative, *i.e.*, if the number of non-quiescent states can increase indefinitely in an evolution process, then another strategy may be employed, such as the length of a tape can increase dynamically. In the 29-state cellular automaton of von Neumann [2], this method is used. There, the blank symbol 0 of a TM's tape is represented by a quiescent state. When the symbol 1 is written on a blank square, then the state of the square is changed to a non-quiescent state. Therefore, even if the initial configuration is finite, a TM in the cellular space can use a potentially infinite tape. In this way, von Neumann made it possible to give a self-reproducing TM in his CA model.

So far, it is not known whether there is a Turing universal reversible PCA with finite configuration in which tape cells of a TM is dynamically created as in the von Neumann's CA. In [68,69], it has been shown that self-reproduction of simple patterns (but having no computing capability) is possible in 2- and 3-dimensional reversible PCAs. Hence, it may be possible to design a mechanism for dynamically creating tape squares in a reversible SPCA. However, many states will be needed to do so in a reversible PCA. Note that, though 81 states in an SPCA is larger than 29 states in von Neumann's CA, the number of local transition rules of the former is much less than that of the latter. Therefore, it is difficult to incorporate various functions, such as universal computing, creating patterns, *etc.*, in a small number of local transition rules. How we can design such a reversible PCA with a small number of states elegantly is the problem.

It is also not known whether there is a Turing universal reversible ESPCA or ETPCA with finite configurations. In some ESPCAs and ETP-CAs, position markers exist, and they are shifted by colliding a glider. However, in them, the shifting direction of a marker and the moving direction of a glider is not parallel, or a glider does not bounce after collision (see, *e.g.*, Fig. 7.10). Therefore, such a marker is not used to implement a counter. If there is a position marker that can be pushed and pulled as in Figs. 8.12 and 8.13), then a counter may be implemented in the ESPCA or ETPCA.

In the RCM M_{exp} in Example 8.2, three counters are used to compute the function $g(n) = 2^n$. We can regard 2^n as a Gödel number of n (see Sec. 4.2.2). Once the number 2^n is obtained, *any* function computed by RCM(k) ($k = 3, 4, \ldots$) can be computed by RCM(2) as described in Lemma 4.5 (see [9,49] for the details). Therefore, for example, the mapping $(2^n, 0) \mapsto (2^{2^n}, 0)$ is computed by RCM(2), which has only *two* counters.

We also discussed T-symmetry of reversible SPCAs. In particular, the Turing universal 81-state reversible SPCA P_3 is shown to be T-symmetric under the involution $H^{\mathrm{rev}} \circ H^{\mathrm{refl}}$ (see Example 8.4). Using this property, an inverse RCM can be obtained from a given RCM by the mechanical operation $H^{\mathrm{rev}} \circ H^{\mathrm{refl}}$.

In the case of ESPCAs and ETPCAs, Turing universal ones shown in Chaps. 6 and 7 are mostly T-symmetric under $H^{\mathrm{rev}} \circ H^{\mathrm{refl}}$ (see Theorem 3.2, Fig. 3.3, and Corollary 3.2). However, ESPCA-02c5df (Sec. 6.3) is strictly T-symmetric, *i.e.*, T-symmetric under H^{rev} (see Theorem 3.1 and Example 3.3). So far, it is not known whether there is a strictly T-symmetric Turing universal 81-state SPCA with finite configurations.

8.8   Exercises

8.8.1   *Paper-and-pencil exercises*

Exercise 8.1. *

(1) Show that a B-turn gadget (Fig. 8.7) is simulated by a pattern composed of R-turn gadgets (Fig. 8.5) in P_3.
(2) Show that a B-turn gadget is simulated by a pattern composed of an RE gadget (Fig. 8.8) and LR-turn gadgets (Fig. 8.4) in P_3.

Exercise 8.2. ** Consider the pattern given in Fig. 8.26 in the 81-state reversible SPCA P_3.

$t = 0$

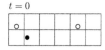

Fig. 8.26   Space-moving pattern in the SPCA P_3.

(1) Show that it is a space-moving pattern.
(2) How is its speed? Note that the speed of a space-moving pattern in P_3 is defined similarly to Definition 2.13.
(3) Show the following: For any small $r > 0$ there is a space-moving pattern in P_3 whose speed is less than cr, where c is the speed of light.

Exercise 8.3. ** Calculate the following numbers of 81-state rotation-symmetric SPCAs.

(1) The total number of all 81-state rotation-symmetric SPCAs.
(2) The total number of reversible ones.

Exercise 8.4.** Sketch a configuration of P_3 that simulates the RCM(2) M'_{half} given in Exercise 4.2.

8.8.2 *Golly exercises*

Exercise 8.5.* Consider the patterns of P_3 obtained in Exercise 8.1. Verify their correctness by simulating them in Golly.

Exercise 8.6.** Consider the problem (3) of Exercise 8.2. Compose various slowly space-moving patterns in P_3 using Golly.

Exercise 8.7.*** Simulate the configuration of RCM(2) M'_{half} designed in Exercise 8.4 using Golly.

Chapter 9

Open Problems and Future Research Problems

In Chaps. 1–8, various problems were posed, and solutions to them were given. They are on the properties of reversible CAs and related reversible systems. A typical problem solved in this book is how we can compose reversible computers in a very simple reversible PCAs. In the course of solving such a problem, we encounter many subproblems. Furthermore, even after successfully finding a solution to the problem, there arises a question whether we can find a different way to solve it.

In this chapter, we show open problems and future research problems. Though there is no clear distinction between open problems and future research problems, the latter ones will need some systematic studies. These problems are not only attractive puzzles, but also clarify essential properties and capabilities of reversible CAs. Of course, there are many problems that are not mentioned below. Readers are encouraged to find new attractive problems, solve them, and further explore the reversible world.

9.1 Open Problems

9.1.1 *Problems of finding patterns with special properties*

9.1.1.1 *Problems on periodic patterns*

Problem 1. Consider the non-conservative ESPCA-0945df (Fig. 5.21), and its periodic pattern of period 60 shown in Fig. 5.24. From this figure, we can see that its maximum diameter is $d_{max} = 8$, and its minimum diameter is $d_{min} = 2$ (see also the pattern file in [21] for Golly: Ch_5_2_1_2_(2)_ESPCA-0945df_periodic.rle). Therefore, their ratio is $d_{max}/d_{min} = 4$. Is there a reversible ESPCA or ETPCA in which a periodic pattern whose maximum diameter d_{max} and minimum diameter d_{min}

satisfies $d_{max}/d_{min} > 4$ exists? Is there an upper bound on the ratio in reversible ESPCAs and ETPCAs?

Note: In the reversible 81-state SPCA P_3 (Sec. 8.1) there is no upper bound on the ratio d_{max}/d_{min} (see Exercise 9.1).

Problem 2. Consider again ESPCA-0945df (Fig. 5.21), and its periodic pattern of period 60 shown in Fig. 5.24. From this figure, we can see that its maximum population is $p_{max} = 34$, and its minimum population is $p_{min} = 2$. Thus, their ratio is $p_{max}/p_{min} = 17$. Is there a reversible non-conservative ESPCA or ETPCA in which a periodic pattern whose maximum population p_{max} and minimum population p_{min} satisfies $p_{max}/p_{min} > 17$ exists? Is there an upper bound of the ratio in reversible ESPCAs and ETPCAs?

9.1.1.2 *Problems on space-moving patterns*

Problem 3. Consider the non-conservative ETPCA-0347 (Fig. 5.35). There is a space-moving pattern called a glider-6 in it (Fig. 5.59). Is there another space-moving pattern in ETPCA-0347 that is essentially different from the glider-6? Here, "essentially different" means that it is not composed of two or more glider-6 patterns chosen from those of $t = 0, \ldots, 5$ in Fig. 5.59.

Problem 4. So far, no space-moving pattern is found in non-conservative reversible ETPCAs except ETPCA-0347. Is there a space-moving pattern in these ETPCAs?

Problem 5. As shown in Sec. 5.3.1.2, each of the non-conservative ESPCAs 094xyf, 098a7f, 098$x$df, 098aef, and 0c4a7f ($x \in \{5, a\}$ and $y \in \{7, b, d, e\}$) has one space-moving pattern, but no other space-moving pattern has been found. Is there a non-conservative ESPCA in which two or more kinds of space-moving patterns exist? Note that they must be essentially different.

Problem 6. Consider the non-conservative ESPCA-0c4a7f (Fig. 5.55), and a glider-15a shown in Fig. 5.56. We can see that its maximum population is $p_{max} = 9$ (at $t = 2$ and 7). Space-moving patterns so far found have rather small maximum population. Is there a reversible ESPCA or ETPCA in which a space-moving pattern having a large maximum population (*i.e.*, ten or more) exists?

9.1.2 *Problems on glider guns*

Problem 7. As shown in Sec. 5.5, glider guns have been found in ESPCAs 094x7f, 094$x$df, 098aef ($x \in \{5, a\}$), and ETPCA-0347. Are there glider guns in other reversible non-conservative ESPCAs or ETPCAs?

Problem 8. The glider gun in ETPCA-0347 given in Sec. 7.1.5 has a special "programmable" feature, *i.e.*, generation of a new glider is controlled by other gliders. Hence, for example, the period of generation is varied. Are there such glider guns in other ESPCAs or ETPCAs?

9.1.3 *Problems on implementing RLEMs*

Problem 9. In Secs. 6.3.3 and 6.4.2 it is shown that an RE is implemented in the reversible ESPCAs 02c5df and 02c5bf using I-gates and I^{-1}-gates. Therefore, adjustment of signal timing is necessary at each gate. Is it possible to implement an RE without using reversible logic gates?

Note: In Secs. 6.1.4 and 6.2.4, an RE is implemented using a position marker that can be shifted by a glider-12. If such a position marker is found in ESPCAs 02c5df and/or 02c5bf, it will become easier to compose an RE. Or, is there any other good method to do so?

Problem 10. In Sec. 7.2.2 it is shown that RLEM 2-2 is implemented in the partial reversible ETPCA-034z, which uses only three local transition rules shown in Fig. 7.25. Is it possible to implement RLEM 2-3, RLRM 2-4, or RLEM 2-17 in ETPCA-034z?

Note: If any one of the three RLEMs is composed, then *any* RLEM can be realized in ETPCA-034z, because of the following reasons. First, RLEM 2-17 is intrinsically universal (Theorem 4.11). Second, RLEMs 2-3 and 2-4 are mutually inverse ones, and they form an intrinsically universal set (Lemma 4.7). Hence, if RLEM 2-3 is implemented in ETPCA-034z, then RLEM 2-4 is also so, and *vice versa*, using a similar method as in Sec. 6.1.3, since ETPCA-0347 is T-symmetric under $H^{\text{rev}} \circ H^{\text{refl}}$.

9.2   Future Research Problems

9.2.1   *Problems on finding universal PCAs*

Problem 11. Find new Turing universal reversible ESPCAs or ETPCAs that have not yet been shown to be universal in this book. In particular, is there a non-conservative reversible ESPCA that is Turing universal?

Note: So far, it has been shown that reversible ESPCAs 01c5ef, 01caef, 02c5df and 02c5bf, reversible ETPCAs 0347, 0157 and 013z, and their duals under reflection and complementation are Turing universal. As observed in Sec. 5.3.1.2, there are several non-conservative ESPCAs that have gliders. Thus, there is a possibility to compose an RLEM in these ESPCAs using these gliders.

Problem 12. Find Turing universal (or fascinating) *irreversible* ESPCAs or ETPCAs.

Note: Although the framework of PCAs has been proposed to design reversible CAs, there may exist interesting irreversible ESPCAs or ETP-CAs. A few examples of irreversible linear PCAs were given in Secs. 5.6.1 and 5.6.2. However, so far, irreversible ones have not yet been fully investigated.

Problem 13. Is there a Turing universal ESPCA or ETPCA with finite configurations?

Note: In Chap. 8, it is shown that any reversible counter machine (RCM) is implemented as a finite configuration in the 81-state reversible PCA P_3. However, it is not known whether such a construction is possible in a reversible ESPCA or ETPCA.

Problem 14. Is there a method for showing Turing-universality of a reversible ESPCA or ETPCA other than the method of using RLEMs or reversible logic gates?

Note: In this book, we used RLEMs rather than reversible logic gates to show Turing universality of several reversible ESPCAs and ETPCAs. By this, RTMs is constructed very simply. However, there may be yet another method of showing Turing universality of them.

9.2.2 Problem on interaction of patterns

Problem 15. Find interesting phenomena in reversible ESPCAs and ETP-CAs by interacting various space-moving and periodic patterns.

Note: There are many ESPCAs and ETPCAs, and also many kinds of space-moving and periodic patterns in each of the PCAs. Therefore, it is difficult to test all possible interactions of the patterns exhaustively. However, we may find interesting or useful phenomena among such interactions. For example, in ETPCA-0347 a large amount of time is required to test all possible cases of collisions of three glider-6's. But, there is an interesting case where one of the three glider-6's is reversibly erased as in Fig. 7.23. It was found from the following two facts: the fact that three gliders are obtained by a collision of two gliders as shown in Fig. 7.22, and the fact that ETPCA is T-symmetric under $H^{rev} \circ H^{refl}$. Thus, we may find further interesting phenomena by a systematic search of collisions of many gliders.

9.2.3 Problem on designing RLEM circuits

Problem 16. Find a simple method of making RTMs out of RLEM 2-2.

Note: As stated in Theorem 4.12, M. Cook and E. Palmiere showed that RLEM 2-2 is Turing universal. However, their method is very complex. Hence, it will be useful if a simpler method is found. Hopefully, RTMs can be simulated in ETPCA-034z using Golly (see Sec. 7.2.2).

9.2.4 Problem on time-reversal symmetry

Problem 17. Many reversible ESPCAs are T-symmetric under the involutions H^{rev}, $H^{rev} \circ H^{refl}$, $H^{rev} \circ H^{comp}$, or $H^{rev} \circ H^{refl} \circ H^{comp}$ (Table 3.1). However, about 42% of the ESPCAs are not T-symmetric under the above four involutions. Is there an involution $\hat{H}$ for each of these ESPCAs, which makes it T-symmetric?

9.2.5 Problem on self-reproduction in reversible PCAs

Problem 18. Can we have a self-reproducing machine in the sense of von Neumann [2] in a reversible cellular space? Namely, is there a reversible SPCA or TPCA in which a self-reproducing pattern with a function of an RTM exists?

Note: It has been shown in [68, 69] that self-reproduction of various objects is possible in 2- and 3-dimensional reversible PCAs. There, self-reproduction is performed by using a description (*i.e.*, "gene") of an object. Thus, it is not a simple self-replication like the Fredkin's CA in Example 1.1. By extending the method, it looks possible to design a reversible PCA in which any RTM can reproduce itself. However, such a PCA may have a huge number of states. The 2-dimensional 5-neighbor PCA given in [68] has 8^5 states. Preferably, the number of states of the reversible PCA is sufficiently small so that a self-reproduction process can be viewed on Golly.

9.3   Exercises

9.3.1   *Paper-and-pencil exercises*

Exercise 9.1.[***] Consider the reversible 81-state SPCA P_3 (Sec. 8.1). Show an outline of a proof of the following statement (see Problem 1). For any $r > 0$, there is a periodic pattern that satisfies $d_{\max}/d_{\min} > r$, where $d_{\max}$ and $d_{\min}$ are maximum and minimum diameters of the periodic pattern in one period, respectively.

Exercise 9.2.[***] Find an attractive open problem or a future research problem, and solve/study it.

Bibliography

[1] A. Trevorrow, T. Rokicki and T. Hutton *et al.*, Golly: an open source, cross-platform application for exploring Conway's Game of Life and other cellular automata, https://golly.sourceforge.io/ (2005).

[2] J. von Neumann, *Theory of Self-reproducing Automata* (ed. A.W. Burks). The University of Illinois Press, Urbana (1966).

[3] S. Ulam, Random processes and transformations, in *Proceedings of the International Congress on Mathematics*, Vol. 2, pp. 264–275 (1952).

[4] M. Gardner, Mathematical games: The fantastic combinations of John Conway's new solitaire game "Life," *Sci. Am.* **223 (4)**, pp. 120–123 (1970), doi:10.1038/scientificamerican1070-120.

[5] M. Gardner, Mathematical games: On cellular automata, self-reproduction, the Garden of Eden and the game "Life," *Sci. Am.* **224 (2)**, pp. 112–117 (1971), doi:10.1038/scientificamerican0271-112.

[6] E. Berlekamp, J. Conway and R. Guy, *Winning Ways for Your Mathematical Plays (2nd Edition)*, Vol. 4. A K Peters, Wellesley, MA (2004), doi:10.1201/9780429487309.

[7] LifeWiki, The wiki for Conway's Game of Life, (2009), https://www.conwaylife.com/wiki/.

[8] N. Johnston and D. Greene, *Conway's Game of Life: Mathematics and Construction*. Self-published (2022), ISBN 978-1-794-81696-1, doi: 10.5281/zenodo.6097284, https://conwaylife.com/book.

[9] K. Morita, *Theory of Reversible Computing*. Springer, Tokyo (2017), doi:10.1007/978-4-431-56606-9.

[10] E. F. Moore, Machine models of self-reproduction, *Mathematical Problems in the Biological Sciences* **14**, pp. 17–33 (1962), doi:10.1090/psapm/014/9961.

[11] C.-G. Langton, Self-reproduction in cellular automata, *Physica D* **10**, pp. 135–144 (1984), doi:10.1016/0167-2789(84)90256-2.

[12] L. Carroll, *The Annotated Alice: Alice's Adventures in Wonderland and Through the Looking Glass* (with an introduction and notes by Martin Gardner). Bramhall House (1960).

[13] J. Myhill, The converse of Moore's Garden-of-Eden theorem, *Proc. Am. Math. Soc.* **14**, pp. 658–686 (1963), doi:10.2307/2034301.

[14] T. Toffoli, Computation and construction universality of reversible cellular automata, *J. Comput. Syst. Sci.* **15**, pp. 213–231 (1977), doi:10.1016/S0022-0000(77)80007-X.

[15] G.-A. Hedlund, Endomorphisms and automorphisms of the shift dynamical system, *Math. Syst. Theory* **3**, pp. 320–375 (1969), doi:10.1007/BF01691062.

[16] D. Richardson, Tessellations with local transformations, *J. Comput. Syst. Sci.* **6**, pp. 373–388 (1972), doi:10.1016/S0022-0000(72)80009-6.

[17] J. Kari, Reversibility and surjectivity problems of cellular automata, *J. Comput. Syst. Sci.* **48**, pp. 149–182 (1994), doi:10.1016/S0022-0000(05)80025-X.

[18] S. Amoroso and Y.-N. Patt, Decision procedures for surjectivity and injectivity of parallel maps for tessellation structures, *J. Comput. Syst. Sci.* **6**, pp. 448–464 (1972), doi:10.1016/S0022-0000(72)80013-8.

[19] K. Sutner, De Bruijn graphs and linear cellular automata, *Complex Syst.* **5**, pp. 19–30 (1991).

[20] N. Margolus, Physics-like model of computation, *Physica D* **10**, pp. 81–95 (1984), doi:10.1016/0167-2789(84)90252-5.

[21] K. Morita, Reversible World of Cellular Automata – Data set for the Golly simulator, and solutions to selected exercises, Hiroshima University Institutional Repository, https://ir.lib.hiroshima-u.ac.jp/00055227 (2024).

[22] K. Morita and M. Harao, Computation universality of one-dimensional reversible (injective) cellular automata, *Trans. IEICE* **E72**, pp. 758–762 (1989), https://ir.lib.hiroshima-u.ac.jp/00048449.

[23] S. Wolfram, *A New Kind of Science.* Wolfram Media Inc. (2002).

[24] K. Morita, Time-reversal symmetries in two-dimensional reversible partitioned cellular automata and their applications, *International J. Parallel Emergent & Distributed Systems* **37**, pp. 479–511 (2022), doi:10.1080/17445760.2022.2102169.

[25] K. Morita and S. Ueno, Computation-universal models of two-dimensional 16-state reversible cellular automata, *IEICE Trans. Inf. & Syst.* **E75-D**, pp. 141–147 (1992), https://ir.lib.hiroshima-u.ac.jp/00048451.

[26] K. Morita, Computing in a simple reversible and conservative cellular automaton, in *Proc. First Asian Symposium on Cellular Automata Technology* (eds. S. Das, G.J. Martinez) AISC 1425, Springer, pp. 3–16 (2022), doi:10.1007/978-981-19-0542-1_1.

[27] K. Morita, Making reversible computing machines in a reversible cellular space, *Bulletin of EATCS* **140**, pp. 41–77 (2023), https://eatcs.org/index.php/on-line-issues.

[28] K. Morita, Composing a rotary element in simple reversible cellular automata to make reversible computers, in *Advances in Cellular Automata* (eds. A. Adamatzky, G.J. Martinez, G.Ch. Sirakoulis). Springer (to appear).

[29] K. Morita, Y. Tojima, K. Imai and T. Ogiro, Universal computing in reversible and number-conserving two-dimensional cellular spaces, in *Collision-based Computing* (ed. A. Adamatzky). Springer, London, pp. 161–199 (2002), doi:10.1007/978-1-4471-0129-1_7.

[30] K. Imai and K. Morita, A computation-universal two-dimensional 8-state triangular reversible cellular automaton, *Theoret. Comput. Sci.* **231**, pp. 181–191 (2000), doi:10.1016/S0304-3975(99)00099-7.

[31] K. Morita, An 8-state simple reversible triangular cellular automaton that exhibits complex behavior, in *AUTOMATA 2016* (eds. M. Cook, T. Neary) LNCS 9664, pp. 170–184 (2016a), doi:10.1007/978-3-319-39300-1_14.

[32] K. Morita, Universality of 8-state reversible and conservative triangular partitioned cellular automata, in *ACRI 2016* (eds. S. El Yacoubi, et al.) LNCS 9863, pp. 45–54 (2016b), doi:10.1007/978-3-319-44365-2_5.

[33] K. Morita, A universal non-conservative reversible elementary triangular partitioned cellular automaton that shows complex behavior, *Natural Computing* **18**, 3, pp. 413–428 (2019), doi:10.1007/s11047-017-9655-9.

[34] J. Lamb and A. Roberts, Time-reversal symmetry in dynamical systems: A survey, *Physica D* **112**, pp. 1–39 (1998), doi:10.1016/S0167-2789(97)00199-1.

[35] A. Gajardo, J. Kari and M. Moreira, On time-symmetry in cellular automata, *J. Comput. Syst. Sci.* **78**, pp. 1115–1126 (2012), doi:10.1016/j.jcss.2012.01.006.

[36] J. Kari, Reversible cellular automata: From fundamental classical results to recent developments, *New Generation Computing* **36**, pp. 145–172 (2018), doi:10.1007/s00354-018-0034-6.

[37] T. Toffoli and N. Margolus, Invertible cellular automata: a review, *Physica D* **45**, pp. 229–253 (1990), doi:10.1016/0167-2789(90)90185-R.

[38] S. Wolfram, *Theory and Applications of Cellular Automata*. World Scientific Publishing (1986).

[39] K. Morita, Time-reversal symmetries in reversible elementary square and triangular partitioned cellular automata, and their data, Hiroshima University Inst. Repository, https://ir.lib.hiroshima-u.ac.jp/00052559 (2022).

[40] A. Church, An unsolvable problem in elementary number theory, *American Journal of Mathematics* **58**, pp. 345–363 (1936), doi:10.2307/2371045.

[41] A.-M. Turing, On computable numbers, with an application to the Entscheidungsproblem, *Proc. London Math. Soc.*, Series 2 **42**, pp. 230–265 (1936).

[42] Y. Lecerf, Machines de Turing réversibles — récursive insolubilité en $n \in \mathbf{N}$ de l'équation $u = \theta^n u$, où θ est un isomorphisme de codes, *Comptes Rendus Hebdomadaires des Séances de L'académie des Sciences* **257**, pp. 2597–2600 (1963).

[43] C.-H. Bennett, Logical reversibility of computation, *IBM J. Res. Dev.* **17**, pp. 525–532 (1973), doi:10.1147/rd.176.0525.

[44] C.-E. Shannon, A universal Turing machine with two internal states, in *Automata Studies*. Princeton University Press, Princeton, NJ, pp. 157–165 (1956).

[45] Y. Rogozhin, Small universal Turing machines, *Theoret. Comput. Sci.* **168**, pp. 215–240 (1996), doi:10.1016/S0304-3975(96)00077-1.

[46] T. Neary and D. Woods, Four small universal Turing machines, *Fundam. Inform.* **91**, pp. 123–144 (2009), doi:10.3233/FI-2009-0036.

[47] M. Cook, Universality in elementary cellular automata, *Complex Syst.* **15**, pp. 1–40 (2004).

[48] M.-L. Minsky, *Computation: Finite and Infinite Machines*. Prentice-Hall, Englewood Cliffs, NJ (1967).

[49] K. Morita, Universality of a reversible two-counter machine, *Theoret. Comput. Sci.* **168**, pp. 303–320 (1996), doi:10.1016/S0304-3975(96)00081-3.

[50] K. Morita, An instruction set for reversible Turing machines, *Acta Informatica* **58**, pp. 377–396 (2021), doi:10.1007/s00236-020-00388-1.

[51] K. Morita, A simple reversible logic element and cellular automata for reversible computing, in *Proc. MCU 2001* (eds. M. Margenstern, Y. Rogozhin), LNCS 2055, pp. 102–113 (2001), doi:10.1007/3-540-45132-3_6.

[52] E. Fredkin and T. Toffoli, Conservative logic, *Int. J. Theoret. Phys.* **21**, pp. 219–253 (1982), doi:10.1007/BF01857727.

[53] Y. Mukai and K. Morita, Realizing reversible logic elements with memory in the billiard ball model, *Int. J. Unconventional Computing* **8**, pp. 47–59 (2012).

[54] K. Morita, T. Ogiro, A. Alhazov and T. Tanizawa, Non-degenerate 2-state reversible logic elements with three or more symbols are all universal, *J. Multiple-Valued Logic and Soft Computing* **18**, pp. 37–54 (2012).

[55] Y. Mukai, T. Ogiro and K. Morita, Universality problems on reversible logic elements with 1-bit memory, *Int. J. Unconventional Computing* **10**, pp. 353–373 (2014).

[56] K. Morita, A new universal logic element for reversible computing, in *Grammars and Automata for String Processing* (eds. C. Martin-Vide, and V. Mitrana). Taylor & Francis, London, pp. 285–294 (2003), doi:10.1201/9780203009642.

[57] J. Lee, F. Peper, S. Adachi and K. Morita, An asynchronous cellular automaton implementing 2-state 2-input 2-output reversed-twin reversible elements, in *Proc. ACRI 2008* (eds. H. Umeo, et al.), LNCS 5191, pp. 67–76 (2008), doi:10.1007/978-3-540-79992-4_9.

[58] N. Ollinger, Universalities in cellular automata, in *Handbook of Natural Computing*. Springer, pp. 189–229 (2012), doi:10.1007/978-3-540-92910-9_6.

[59] K. Morita and T. Ogiro, How can we construct reversible machines out of reversible logic element with memory? in *Computing with New Resources* (eds. C.S. Calude et al.), LNCS 8808. Springer, pp. 352–366 (2014), doi:10.1007/978-3-319-13350-8_26.

[60] M. Ito, N. Osato and M. Nasu, Linear cellular automata over $\mathbf{Z}_m$, *J. Computer and System Sciences* **27**, pp. 125–140 (1983), doi:10.1016/0022-0000(83)90033-8.

[61] G.-J. Martinez, H.-V. McIntosh and J.-C. Seck-Tuoh-Mora, Gliders in rule 110, *Int. J. Unconventional Computing* **2**, pp. 1–49 (2005).

[62] A. Adamatzky, and G.-J Martinez (eds.), *Designing Beauty: The Art of Cellular Automata*. Springer, Cham (2016), doi:10.1007/978-3-319-27270-2.

[63] K. Morita, Constructing reversible Turing machines in a reversible and conservative elementary triangular cellular automaton, *J. Automata, Languages and Combinatorics* **26**, pp. 125–144 (2021), doi:10.25596/jalc-2021-125.

[64] K. Morita, Gliders in the Game of Life and in a reversible cellular automaton, in *The Mathematical Artist: A Tribute To John Horton Conway* (eds. S. Das, S. Roy, K. Bhattacharjee). Springer, Cham, pp. 105–138 (2022), doi:10.1007/978-3-031-03986-7_5.

[65] K. Morita, Finding a pathway from reversible microscopic laws to reversible computers, *Int. J. Unconventional Computing* **13**, pp. 203–213 (2017).

[66] K. Morita, How can we construct reversible Turing machines in a very simple reversible cellular automaton?, in *Proc. RC 2021* (eds. S. Yamashita, T. Yokoyama) LNCS 12805, Springer, pp. 3–21 (2021), doi:10.1007/978-3-030-79837-6_1.

[67] K. Morita, Y. Tojima and K. Imai, A simple computer embedded in a reversible and number-conserving two-dimensional cellular space, *Multiple-Valued Logic* **6**, 5-6, pp. 483–514 (2001).

[68] K. Morita and K. Imai, Self-reproduction in a reversible cellular space, *Theoret. Comput. Sci.* **168**, pp. 337–366 (1996), doi:10.1016/S0304-3975(96)00083-7.

[69] K. Imai, T. Hori and K. Morita, Self-reproduction in three-dimensional reversible cellular space, *Artificial Life* **8**, pp. 155–174 (2002), doi:10.1162/106454602320184220.

Index

Printed in the United States
by Baker & Taylor Publisher Services